SUSTAINABILITY AND CHANGE IN RURAL AUSTRALIA

CHRIS COCKLIN is Professor of Geography and Environmental Science at Monash University. He has written on sustainability in a rural context, land use change, rural communities and environmental issues. He is a member of the International Geographical Union Commission on Sustainable Rural Systems and recently edited, with Ian Bowler and Chris Bryant, *The Sustainability of Rural Systems* (Kluwer, 2002).

JACQUI DIBDEN is a social anthropologist who has carried out research in Indonesia and Australia. She also spent many years working in community development in Australian country areas. At Monash University, she has undertaken research on social capability of land managers, rural restructuring, sustainability of rural towns and (with Chris Cocklin) the impact of deregulation and other changes on the dairy industry in Australia.

SUSTAINABILITY AND CHANGE IN RURAL AUSTRALIA

Macquarie
Regional Library

Edited by
Chris Cocklin and Jacqui Dibden

UNSW
PRESS

A UNSW PRESS BOOK

Published by
University of New South Wales Press Ltd
University of New South Wales
Sydney NSW 2052
AUSTRALIA
www.unswpress.com.au

ACADEMY OF THE SOCIAL
SCIENCES IN AUSTRALIA

This book is a special project of the Academy
of the Social Sciences in Australia, funded by
the Australian Research Council.

National Library of Australia
Cataloguing-in-Publication entry:

Sustainability and change in rural Australia.
    Includes index.
    ISBN 0 86840 631 7.

    1. Sustainable development – Australia. 2. Rural population –
    Australia. 3. Rural development – Australia. 4. Australia – Rural
    conditions. I. Cocklin, Chris. II. Dibden, Jacqui. III. Academy
    of the Social Sciences in Australia.

    307.720994

Printer dbooks
Cover design Di Quick
Cover image PhotoLibrary.com
Layout Lamond Art & Design

CONTENTS

PREFACE

In 2000, the Academy of the Social Sciences in Australia was awarded a grant by the Australian Research Council, under its Learned Academies Special Projects Scheme, to carry out research on 'The Sustainability of Australian Rural Communities'. Six university-based teams from across Australia convened in Canberra in February 2001 to develop a shared conceptual framework and to outline a methodology for the research program. Subsequently, over several months during 2001, each of the teams undertook an investigation of sustainability in the context of a particular rural community – Narrogin in Western Australia, the Gilbert Valley in South Australia, 'Tarra' in Victoria, Tumbarumba and Guyra in New South Wales, and Monto in Queensland. Each of these studies was guided by the agreed conceptual framework and methodology. The six case studies were published in 2003 under the title *Community Sustainability in Rural Australia: A Question of Capital?* Collectively, the case studies offer insight into the difficult circumstances that these small communities face, but they also point to the various factors that have enabled them to survive in the face of their many challenges.

In this second volume arising from the project, the authors have taken a more thematic approach to issues associated with the sustainability of rural communities. The chapters consider the state of rural communities and the main dimensions of sustainability, and review both the tried and prospective interventions to improve community sustainability. To an extent, each of the chapters draws on the six case studies of community sustainability that have underpinned the project.

In addition to the project research teams, whose members are represented through the authorship of the individual chapters in this book, the Academy of the Social Sciences in Australia was involved through a Project Steering Committee. The members of this committee were Professor Sue Richardson (Flinders University), Professor Lois Bryson (RMIT) and Professor Jim Walmsley (University of New England). Dr John Robertson, Research Director of the Academy of the Social Sciences in Australia, co-ordinated the administrative and logistical support to the project. The valuable intellectual input of all of these people to the project is gratefully acknowledged. We should also like to thank Professor Graeme Davison (Monash University), who first proposed this project, and Lesley Sutherland, who provided invaluable assistance with the final editing process. The financial support to the project provided by the Australian Research Council is gratefully acknowledged.

Chris Cocklin and Jacqui Dibden
MELBOURNE

CONTRIBUTORS

Margaret Alston is Professor of Social Work and Human Services and is Director of the Centre for Rural Social Research at Charles Sturt University. She has published widely in the field of rural gender and rural social issues. She completed a study titled *Breaking through the Grass Ceiling: Women, Power and Leadership in Rural Organisations* (Harwood Publishers, 2000), funded by the Australian Research Council. Recent research projects include social impacts of drought, rural issues, rural women and sport, violence against women in rural areas, rural youth and employment, and women and agriculture.

Neil Argent is a Senior Lecturer in Human Geography at the University of New England (UNE). Prior to joining UNE in 1996, he worked variously as a shearer, rouseabout, farm hand and small grazier in the Barossa Ranges of South Australia. His research interests centre on rural social and demographic change and service provision issues in rural areas. He recently co-wrote a chapter on rural bank branch closures with Fran Rolley (UNE) for *Land of Discontent* (edited by B Pritchard & P McManus, UNSW Press, 2002). His research has been published in *Australian Geographer*, *Australian Geographical Studies*, *Journal of Rural Studies* and *Geoforum*.

Alan Black was Foundation Professor of Sociology/Anthropology at Edith Cowan University, Perth, before his retirement at the end of 2003. As an Emeritus Professor, he retains a link with the Edith Cowan University Centre for Social Research, which he established in 1997. Many of his research projects relate to aspects of community capacity, sustainability and wellbeing, especially in rural and regional Australia. He and his colleagues have recently completed a large national survey on 'Wellbeing and Security in Australia'. He is also the representative for Australia on the International Advisory Council for the World Values Survey.

Lynda Cheshire is a Lecturer in Sociology with the School of Social Science at the University of Queensland. Her research interests are in the areas of rural and regional governance, rural development, local responses to change and restructuring, and governmentality theory. She is currently working on a number of projects on state–community–industry engagement practices in regional Queensland. Her book, *Governing Rural Development: Discourses and Practices of Self-help in Australian Rural Policy* (Ashgate), will be published in late 2004.

Chris Cocklin is Professor of Geography and Environmental Science at Monash University, Melbourne. He has written widely on sustainability in a rural context, land use change, rural communities and environmental issues. He recently edited, with Ian Bowler and Chris Bryant, *The Sustainability of Rural Systems* (Kluwer, 2002). He was the Project Director of the Academy of the Social Sciences in Australia project, on which the present book is based, and co-edited the previous project publication, *Community Sustainability in Rural Australia: A Question of Capital?* (Centre for Rural Social Research, Charles Sturt University, 2003).

Graeme Davison teaches history at Monash University, where he is a Sir John Monash Distinguished Professor. His main field of interest is in the history of cities, where his publications include *The Rise and Fall of Marvellous Melbourne* (Melbourne University Press, 1978) and *Car Wars: How the Car Won our Hearts and Conquered our Cities* (Allen & Unwin, 2004), and in Australian public history, the subject of *The Use and Abuse of Australian History* (Allen & Unwin, 2000). He is a co-editor of the *Oxford Companion to Australian History* (Oxford University Press, 1998).

Jacqui Dibden is a Research Fellow with the Monash Regional Australia Project (MRAP), Monash University. She has a particular interest in social change and has carried out ethnographic research in Indonesia and Australia. Jacqui spent many years working in community development in country areas. At Monash University, she has undertaken research on social capability of land managers, rural restructuring, sustainability of rural towns and (with Chris Cocklin) the impact of deregulation and other changes on the dairy industry in Australia. She co-edited *All Change! Gippsland Perspectives on Regional Australia in Transition* (MRAP, 2001). She has also co-authored a number of book chapters and journal articles with Chris Cocklin and others.

Ian Gray is Associate Director of the Centre for Rural Social Research (CRSR) at Charles Sturt University, Wagga Wagga. His research projects have covered many issues related to the sustainability of farms and towns. With Professor Geoffrey Lawrence, he wrote *A Future for Regional Australia* (Cambridge University Press, 2001). He has also published in Australian and international journals, including *The Australian Journal of Social Issues*, *Sociologia Ruralis* and the *Journal of the Community Development Society*. His other books include *Politics in Place: Social Power Relations in an Australian Country Town* (Cambridge University Press, 1991) and, with other authors, *Immigrant Settlement in Country Areas* (AGPS, 1991), *Coping with Change: Australian Farmers in the 1990s* (CRSR, 1993) and *Australian Farm Families' Experience of Drought* (RIRDC, 1999). His research interests cover sociology of community, rural society, local government, urban society, transportation and the environment.

Trevor Griffin is an independent researcher based at the University of Adelaide, where he previously held a Senior Lectureship in Geography. His broad interests in both human and physical geography are allied to specialist interests in cartography, GIS and statistical analysis. He was joint editor, with Murray McCaskill, of the *Atlas of South Australia* (South Australian Government Printing Division in association with Wakefield Press, 1986).

Graeme Hugo is Federation Fellow, Professor of the Department of Geographical and Environmental Studies and Director of the National Centre for Social Applications of Geographical Information Systems at the University of Adelaide. He is the author of over 200 books, articles in scholarly journals and chapters in books, as well as a large number of conference papers and reports. His books include *Australia's Changing Population* (Oxford University Press, 1986), *Worlds in Motion: Understanding International Migration at Century's End* (with DS Massey, J Arango, A Kouaouci, A Pellegrino & JE Taylor, Oxford University Press, 1998), several of the 1986–96 census-based *Atlas of the Australian People* series (AGPS) and *Australian Immigration: A Survey of the Issues* (with Mark Wooden, Robert Holton & Judith Sloan, AGPS, 1989–96). In 2002, he secured a $1.125 million ARC Federation Fellowship over five years for his research project, 'The new paradigm of international migration to and from Australia: dimensions, causes and implications'.

Geoffrey Lawrence is Professor of Sociology, and Head of the School of Social Science, at the University of Queensland. He has had extensive experience in rural social research in Australia, completing many studies in the areas of family farming, agrifood networks, and the social aspects of biotechnology. His most recent co-authored/co-edited books are *Recoding Nature: Critical Perspectives on Genetic Engineering* (UNSW Press, 2004), *Globalization, Localization and Sustainable Livelihoods* (Ashgate, 2003) and *A Future for Regional Australia: Escaping Global Misfortune* (Cambridge University Press, 2001).

Peter Smailes is a Visiting Research Fellow at the University of Adelaide, having recently retired from a Senior Lectureship in Geography, which he occupied at the same university from 1979 to 2001. He has previously held academic positions at the University of New England, Flinders University and the University of Oslo, Norway. His major research field at all these institutions has been rural social geography, including the impacts of migration, demographic change, accessibility and rural service provision on the viability of rural communities.

Richard Stayner is Principal Project Director in the Institute for Rural Futures at the University of New England in Armidale, New South Wales. He has previously held teaching and research positions at the University of Queensland, the University of Newcastle-upon-Tyne (UK) and Monash University. Over the past 20 years, his major research interests have been in the economics of agriculture, in particular structural adjustment and the way in which social factors influence the adjustment behaviour of farmers, and the economics of rural communities and regions. He has also conducted a number of consultancies on local and regional economic development.

Matthew Tonts is a Lecturer in Geography at the University of Western Australia. He has interests in regional development policy in Australia and Europe, rural social and economic change, and rural planning. His current work includes research on the social geography of sport in rural Australia, the role of corporate agribusiness and regional development, and planning issues associated with the plantation forestry industry. The findings of this research have appeared in a number of international journals, including the *Journal of Rural Studies, Regional Studies* and *Tijdschrift voor Economische en Sociale Geografie*.

1

INTRODUCTION

Jacqui Dibden and Chris Cocklin

In Australia today, country people are constantly confronted with the reality of change, but they are also exhorted to 'embrace change' and move on to a better future – or at least, ensure that they are the lucky ones who manage to survive. Opposing this view of the inexorable march of progress are those who yearn for a return to a remembered past when rural towns and farms were less beleaguered and formed the core of the nation's identity. As Graeme Davison shows in chapter 3, this past is largely an illusion – farmers have always struggled and country places have waxed and waned – but it is true that rural communities enjoyed a measure of support and a belief in their value that in recent years has diminished. At the same time, environmental problems associated with agriculture and human settlement have achieved a gravity that makes it apparent that hard choices must be made about the future of the countryside.

Much of the discussion about a future for rural Australia revolves around questions of sustainability. Is farming sustainable or can it be made so? Are rural towns sustainable or will some of them inevitably disappear? Are the natural environment and the ecosystem services that it provides sustainable or do we face a future of dwindling water supplies, spreading salinisation and degraded landscapes? What are the repercussions of declining rural sustainability for the country as a whole? The salience of these questions has been demonstrated both by a rural backlash, shown through electoral reversals and the rise of new parties, and by the growing evidence of the ramifications of rural environmental problems for country people and indeed for the entire Australian population.

The sustainability of rural communities in Australia was the focus of a multi-university project in 2000–02. The objective of this project, carried out under the auspices of the Academy of the Social Sciences in Australia, was to explore the factors that contribute to – or undermine – the sustainability of rural communities in Australia. To tackle this task, research groups from six universities came together to debate the major issues and develop a common conceptual framework. Agreement was reached on a set of ideas about sustainability, including the notion that a number of 'capitals' may be seen as underpinning sustainability. Each team then endeavoured to apply this framework within a case study of a selected Australian rural 'community'. Although the meanings associated with concept of community have been much debated, it has generally been used in two main ways: to refer to 'communities of place' (or 'communities of location') and to 'communities of interest'. The Academy of the Social Sciences in Australia project approached community sustainability from a place-based perspective, although inevitably at times the case studies referred to communities of interest, which often extend beyond the defined spatial boundaries of particular localities. We recognised that local rural communities are often characterised as much by differences – for example, between farmers and non-farmers – as by shared interests, and that 'diverse interest groups can subscribe to a shared set of symbols and imagine themselves as a community, yet attribute different meanings to these symbols' (Scott et al., 2000: 438).

SUSTAINABILITY

'Sustainability', like 'community', is an ambiguous and contested concept. Both concepts have generated an extensive literature that has struggled with 'the conundrums of definition' (Cocklin & Alston, 2003: 3).[1] In relation to sustainability, Cocklin and Alston (2003: 3) point out that:

> ... while some people have lamented the lack of consensus on definition, to do so overlooks the fact that communities and societies will always contest the relative emphasis that is given to economic growth, social justice, and environmental protection. It should therefore be taken as given that sustainability will be subject to mediation through contest and debate. The objective then shifts from a fruitless search for consensual meaning, to one of under-

standing the many and often competing perspectives on sustainability and to understanding how these contests play out in social space – the role of actors, the multiple constructions of meaning, power relationships, the authority of competing knowledges and discourses, and the implications of the many and varied pathways towards sustainability.

One influential set of definitions of sustainability stems from the international agreements that Australia has entered into, which are outlined by Alan Black, in chapter 2, and by Matthew Tonts, in chapter 11. For the purposes of this collaborative Academy of the Social Sciences in Australia project, the following points were acknowledged:

- Sustainability involves balance and compromise among social, economic and environmental priorities, or what is now popularly referred to as the 'triple bottom line'.
- Sustainability refers to progress towards (or retrogression from) preferred futures, rather than to a defined endpoint. This means that sustainability is not an absolute concept, but a relative one.
- Definitions and interpretations of sustainability are socially contested, because people assign different priorities to social, economic and environmental assets and outcomes.
- Some indicators may point to communities that are progressing towards or retrogressing from sustainability (for example, rising/falling incomes, increasing/decreasing population, increasing/decreasing employment), but these indicators are not consistently reliable markers of direction.
- Sustainability implies equity, both within contemporary society (intragenerational) and in terms of the legacy for future generations (intergenerational) (Cocklin & Alston, 2003: 3–4).

CAPITALS

The concept of 'capitals' was adopted by the Academy of the Social Sciences in Australia project as a framework within which to categorise and (where possible) measure community and social change, and to evaluate the sustainability of each of the case-study communities. The starting proposition for our research was that sustainability might be assessed in relation to the growth, maintenance or decline in the various stocks of 'capital' available; for example, a decline in stocks of capital could be seen as a signal that a community or region might not be 'sustainable'. It was decided that our focus would be not only on the stocks of the various resources but also on the flows of goods and services generated by them. Drawing on the extensive literature on sustainability, five 'capitals' were defined – natural, human, social, institutional and produced.

NATURAL CAPITAL

In relation to sustainable communities, Hart (2000) identifies three types of natural capital: natural resources, ecosystem services, and the aesthetics or beauty of nature. Natural resources are those things – such as water, plants, animals, minerals and fossil fuels – that we can take from the natural environment and use either in their unmodified form or transformed through production processes. Ecosystem services are natural processes from which humans (and other species) benefit in some way; for example, the filtering and absorption of floodwaters by wetlands, and the production of oxygen and sequestration of carbon by growing plants. Ecosystem services have also been defined (Cocklin et al., 2003: 32) as:

> ... those public good services, which generally come from natural areas, but which can also result from sustainable management of land and water. Included are the provision of clean air and water, biodiversity services and sequestration of carbon.

This definition is interesting in that it allows for human agency in production of environmental services. The third aspect of natural capital is the beauty of the natural environment; for example, trees and flowers, rivers, lakes, mountains and coastal scenery. The first and third types of natural capital represent different and often conflicting valuations and uses of aspects of the natural environment, resulting in disagreements and debate about the appropriate 'balance between exploitation and conservation' (Sorensen, 2000: 11).

HUMAN CAPITAL

The abilities, knowledge and skills of individuals are seen as making up the human capital of a community. In discussions of sustainability, emphasis is placed on:

> ... the capacity [of individuals] to contribute through production, decision-making, social interaction, innovation, and in other ways to

a community. The ability to use technology (but not the technology itself) is also included in human capital, as technology extends the abilities of people in many directions. Likewise, the ability to discover, develop or use new knowledge and skills is a vital aspect of human capital. (Cocklin & Alston, 2003: 4)

Individual attributes that are seen as particularly valuable for achieving community viability – often in combination with social capital – are leadership, ability to solve problems and commitment to the locality.

SOCIAL CAPITAL

Relationships between people linked in various ways constitute the social capital of a community (or other entity). Social capital is often seen as taking two forms: 'bonding social capital', or relationships between relatives and close friends or within a social group, and 'bridging social capital', which refers to relationships with acquaintances or with people from other social groups that may differ in religion, ethnicity, gender or socioeconomic status (Woolcock & Narayan, 2000). Associated with these relationships or social networks are various other attributes, such as trust, reciprocity and shared norms and values, which maintain the ties between individuals and enable common action. To these attributes, Black and Hughes (2001) add altruism, shared beliefs, tolerance, a sense of belonging to a community, self-reliance and self-help. Social capital occupies a privileged position in discussions of sustainability, since it has been seen as facilitating the creation or use of the other types of capital. It is considered to underpin the ability of a community or group to act as (and in the interests of) a community, and to identify and work towards community-based outcomes. The role of human agency in achieving community sustainability is underlined by the distinction made in some of the social capital literature between capital (the stock of social resources) and capacity (the ability to draw on capital for valued purposes).

INSTITUTIONAL CAPITAL

This term refers to the institutional structures and mechanisms present in a community. The three main types of institutional structures generally identified are:

- the public sector, in Australia made up of the institutions and agencies of federal, state and local government

- the private enterprise sector, comprising non-government enterprises producing goods and services for profit, together with the market mechanisms through which goods and services are exchanged
- the 'third' sector made up of non-governmental, not-for-profit organisations and institutions

The distinction between social capital and institutional capital is often blurred in practice, since the effective functioning of social institutions also depends, to some extent, on the existence of relationships of trust, co-operation and reciprocity, although these are generally less personal in nature.

PRODUCED CAPITAL

This category – sometimes referred to as 'economic capital' – is made up of harvested or manufactured products, the built environment (for example, buildings, roads, railways, bridges, communication systems and reticulation systems for energy and water), and financial resources, such as money. In addition, both cultural and intellectual property have also been considered forms of produced economic capital. In evaluating the stocks of economic capital, both assets and liabilities need to be taken into account.

The term 'capital' is often used only in a positive sense. However, two provisos must be made. Firstly, whereas the development and sustainability literature emphasises the use of capitals for collective action, an alternative view (associated with Pierre Bourdieu) regards valued resources as 'objects of struggle' between individuals and families (Swartz, 1997: 43). The extent to which capitals represent individual or familial assets rather than community resources is likely to vary within different local contexts and situations. Secondly, social capital has been identified as having a 'dark side', in that mutually supportive interpersonal relationships within a network or community may be accompanied by prejudice or hostility towards outsiders. Other 'capitals' might also involve a 'dark side'. In this light, it is more appropriate to regard capitals as neither inherently positive nor negative, but as capable of being used in ways that can have either beneficial or adverse consequences, or a mixture of both. For example, the productive use of land and water in agriculture or industry may have economic benefits but result in pollution, with damaging effects on both the natural environment and people's health (a component of human capital).

SUSTAINABILITY IN THE MIDST OF CHANGE

Sustainability does not imply stasis. As Alan Black notes in chapter 2, one useful definition of the attributes of a sustainable society includes flexibility, or the ability to 'change in order to respond to various contingencies'. These contingencies may include shifts in the capitals contributing to sustainability. One proposal, outlined by Black in chapter 2, revolves around the possibility of substituting reproducible assets (produced capital) for exhaustible stocks of natural capital. International definitions of sustainability (see chapters 2 and 11) assert the principles of intergenerational equity, whereby the present generation endeavours to pass on to future generations a composite stock of capital at least equal to that which currently exists. However, as Tonts and Black (2003: 108) observe, this raises the question of determining what combination of resources, or capitals, are passed on to the next generation. Rather than viewing the capitals present within an area as a random assemblage, there may be key combinations required for sustainability, with one form of capital depending on the presence, at particular levels, of other capitals. For example, the full value of human or social capital may only be realised in the presence of 'sufficient' levels of natural and produced capital. As many Australian rural communities are discovering, it may become difficult to use these capitals effectively if the land on which they depend has become degraded by salinity. More optimistically, Tonts and Black (2003: 132) suggest that 'the arrest and reversal of salinity is likely to require the successful deployment of each of these forms of capital – institutional, social, human, and produced'.

If we accept that realising the value of some types of capital may be contingent on the existence of others, this also raises the question of whether some of the capitals are more important than others. In this project, social and natural capital have been given a prominent role, with natural capital frequently seen as threatened and posing limits to sustainability, while social capital provides the means by which agency is applied to issues of sustainability. Social capital has been called upon to solve problems relating to viability of rural communities or the need to undertake environmental remediation over larger areas than the individual property. As Dibden and Cocklin (2003: 173) pointed out:

> The networks of reciprocity and trust which constitute social capital provide the possibility of collective action, enabling a shift from

capitals seen as resources for individuals to resources which – at least potentially – may be used for community benefit or the 'public good'.

Similarly, Smailes and Hugo (2003: 98–99) suggest that 'social capital is the most important influence on the sustainability of rural communities' and so 'the maintenance of social capital is a particularly vital ingredient in sustainability'.

A difficult question posed by the Academy of the Social Sciences in Australia project was how to determine the significance and consequences of the decline or growth of various capitals for the sustainability of a community, a region, or the rural sector. Thus:

> ... while some forms of capital may be declining, becoming obsolescent, or used up, other forms may be accumulating – the human capital of blacksmiths, for example, was ultimately replaced by that of motor mechanics. This reminds us that development both creates and destroys some forms of capital. (Cocklin & Alston, 2003: 5)

Similarly, as several chapters in this book reveal, changes in population mobility and the demographic profile of rural areas may have mixed impacts on the social and human capital of the locality. Attempts to determine whether the net change has been positive or negative run up against the difficulty of establishing a single measuring rod. As Stayner (2003: 39) argued, in his account of the town of Guyra:

> The multiplicity of capital items makes it difficult to estimate aggregate quantitative measures of total capital, except perhaps for produced capital, where the values of capital assets of public and private sector enterprises are represented in their financial accounts. Some types of capital are intangible, and their existence is only evident through the services they generate.

This problem of measurement has implications not only for the assessment of the sustainability of any particular community or place, but also for comparisons between communities and localities.

Determining how much capital, of the various kinds, we need in order to achieve sustainability is a question of judgment, involving compromise and trade-offs among social, economic and environmental outcomes. As the Academy of the Social Sciences in Australia project revealed, the value of the capitals-based approach to the case-study research was that it provided a systematic framework for assessing the extent to which the conditions within a locality might favour or inhibit sustainability (variously defined), but it was unable to provide an unequivocal verdict on the sustainability or otherwise of the communities studied.

Moving beyond problems of measurement or weighting of the various capitals, fundamental questions arise about the nature of sustainability and what it is we are trying to sustain. Two overarching issues are relevant to the evaluation of community sustainability. One is geographic context, in particular the relationship between a specific place or social catchment (Hugo, 2001) and other towns in the settlement network. An important point arising from the Academy of the Social Sciences in Australia project, and from other rural studies is:

> ... that towns and communities do not exist and operate independently of other places in the settlement network, and that the fortunes of any one town are partially contingent on the fortunes of its near neighbours. (Cocklin & Alston, 2003: 6)

In a situation in which rural communities are encouraged to be self-reliant and competitive, there is a perceived risk that an improvement in the condition of one settlement may threaten the sustainability of neighbouring towns. And indeed, as chapters 4 and 5 show, there are striking divergences in the patterns of population growth and decline in different areas. The tendency for both public and private services to become increasingly concentrated in larger service centres has, as Richard Stayner demonstrates in chapter 7, resulted in business, employment and financial resources leaking away from smaller towns, and people have tended to follow. Location has a significant bearing on the possibilities for community viability, with proximity to major centres, availability and condition of natural resources, and the amenity of the built and natural environment all playing an important part.

In considering sustainability, there is a tendency to focus on the internal (or endogenous) characteristics of a particular community or region – a tendency promoted by the contemporary policy discourse of self-reliance. However, a second overarching issue identified by our study is the need to consider the exogenous factors that influence community sustainability and the extent to which a particular locality is prospectively impacted by influences operating at other levels. These influences include government policies (for example, in relation to service provision and environmental regulation), commodity prices (which in turn are shaped by shifts in consumer sentiment and international exchange rates), and regional or national environmental agendas. The impact of globalisation on local communities – the 'global/local nexus' (see, for example, Lobao, 1996; McMichael, 1996) – is discussed by Geoffrey Lawrence in chapter 6, while Matthew Tonts, in chapter 11, considers the role of government in promoting sustainability. The interaction between the 'local' and the 'external' (Bebbington & Perreault, 1999: 415) may produce adverse effects; for example, local social capital may be undermined by the dismantling of wider-scale social supports, but this interaction also provides the possibility for a community to access capitals located at wider scales. Stayner (2003: 63) concluded from his case study that 'the stocks of capital that are necessary to ensure the sustainability of community functions in Guyra can not all be generated from within the community'. This understanding leads Smailes and Hugo (2003) to raise the important conceptual question of whether the capitals are attributes of the communities themselves, or of the wider milieux of which they are a part. Indeed, the identification of people with only one community is somewhat artificial. As Alan Black points out in chapter 2, 'people may simultaneously be part of several communities defined at different geographical scales'.

In assessing sustainability, historical as well as geographical context is important. As Graeme Davison shows in chapter 3, rural towns have gone through periods of both expansion and decline. The Academy of the Social Sciences in Australia project case studies were essentially 'snapshots in time', although each traced the major historical events that shaped the community and acknowledged the historical influences on the present stock of capitals (Cocklin & Alston, 2003: 6). The short time perspective employed in most studies of contemporary rural communities makes it difficult to assess how sustainable a community is at the present time, let alone its prospects for the future. As Jacqui Dibden and Lynda Cheshire point out in chapter 12, the success of community development efforts may be ephemeral in a climate in which numerous small towns are obliged to compete for a floating population of tourists and new residents. Sustainability, by definition, requires a very long time perspective, spanning generations and stretching into the indefinite future.

The final question addressed in our study is also the most difficult. This question is: sustainability for whom, or for what? This leads us back to the problem of achieving a balance between the three dimensions of sustainability. If greater weight is given to the health of the environment, the decline of rural communities may be regarded as acceptable in the wider scheme of things. Even where the focus is drawn in to a more human-centred vision of sustainability, questions arise about who will benefit from initiatives to promote sustainability, and who is driving the agenda? The definitions of sustainability stemming from the Brundtland Report (see chapters 2 and 11) emphasise not only intergenerational but also intragenerational equity, which relates to the distribution of capital within a community, and the extent to which marginalised groups benefit from any improvement in the aggregate level of community sustainability. Despite the tendency to view rural communities as homogeneous and unified by common interests, many are in fact characterised by socio-demographic heterogeneity and disparities in power and influence. As a result, initiatives in support of sustainability may not only be devised in accordance with the priorities of the powerful, but they may also – often inadvertently – be the major beneficiaries.

STRUCTURE OF THIS BOOK

This book confronts the issues of sustainability and change in rural Australia from a variety of different directions and a range of disciplinary perspectives. All are grounded in the same background of participation in the Academy of the Social Sciences in Australia project with its 'capitals' framework. As a result, the various capitals make their appearance again in the chapters that follow, although with varying degrees of emphasis. The book is organised into three main parts. The first sets out the conceptual foundations and provides both a historical and a contemporary overview of the state of rural communities. The second part of the book analyses the three main dimensions of sustainability – economic, social and environmental – in relation to contemporary processes. The third part of the book considers interventions at various levels (government, local community, individual) in terms of improving the sustainability of rural communities.

PART I SUSTAINABILITY IN CONTEXT

Because the terms 'community', 'rural' and 'sustainability' are each contestable, Alan Black begins chapter 2 with an exploration of the scope of these terms. He then sets out the dimensions of sustainability – ecological, economic and social – and discusses the ways in which the concept 'has been applied to a wide variety of phenomena, such as social practices, farming systems, resource use, economies, communities and societies'. He examines in detail the use of the term in relation to communities, to forms of development and to livelihoods. Black sees sustainability of rural communities as problematic, both because of the difficulty of reaching agreement on what constitutes sustainability and because of the catastrophic events ('shocks') and wider trends ('stresses') that impinge on rural communities.

Graeme Davison, in chapter 3, provides a valuable corrective to commonly held beliefs that conditions in rural Australia are both worse and radically different from the past. While the time span for most of the other chapters is the past three decades – the period within which 'economic rationalism' (the Australian version of neoliberalism) became the established orthodoxy – Davison provides the longer perspective required to fully understand what is new and what is a recurrent feature of the Australian rural condition. As he notes with reference to recent studies of 'rural discontent': 'Their framework of reference is often short, just a decade or so; yet the problems themselves are often long-term and intractable.' A century ago, young country people were already moving to the cities, resources were becoming degraded and rural towns were in decline. Thus, as Davison observes, 'sustainability is not only an economic, demographic and environmental concept; it is also a historical concept. It links future expectations and present policies with past experience'. However, although many of the trends identified in recent decades were already present in earlier periods, there has been a marked change in attitudes towards the countryside and the prospects for its development. Davison notes:

> Sustainability, with its assumptions of maintenance rather than growth, and of an environmental, rather than economic or political, frame of reference, represents a belated climb-down from the more sanguine expectations of earlier generations.

Chapters 4 and 5 present contrasting views on socio-demographic changes within rural Australia, examined at different scales. Both are agreed in seeing non-metropolitan Australia as a dynamic region, but differ in the weight given to a posited polarisation between areas of growth and decline. Graeme Hugo, in chapter 4, seeks to expand our knowledge of the people of regional Australia – who, contrary to common preconceptions, represent a third of the population – and to dispel the stereotypes of 'non-metropolitan Australian communities and their residents ... as being less dynamic and less differentiated than the nation's metropolitan sector'. He argues that the increased diversity of Australia's non-metropolitan population offers considerable potential for achievement of sustainability in rural Australia. However, Hugo also argues that, along with increased diversity in non-metropolitan Australia, there has been a polarisation within the sector, which 'presents major challenges to policy-makers'.

Peter Smailes, Trevor Griffin and Neil Argent, in chapter 5, take Hugo's broad indications of increasing heterogeneity among, and divergence between, place-based communities as hypotheses to be tested. In this chapter, the focus changes from the broad national overview presented in chapter 4 to a closer examination of rural communities within a crescent-shaped area in south-eastern Australia. The study explores the extent of change in the broad social environment over the past two decades (1981–2001); that is, the period of very rapid restructuring following the radical economic changes commenced by the Hawke and Keating governments, and the adoption of neoliberal policies by the succeeding Howard governments. The analysis reveals that, while sharp contrasts are apparent between the inland regions and the eastern coastal belt, 'the remaining rural communities do not fit conveniently into either of these polarised categories'. Part of the differentiation in apparent sustainability of rural communities relates to two non-productivist factors – 'amenity' and 'accessibility/remoteness'.

PART II ISSUES IN CONTEMPORARY RURAL AUSTRALIA

Geoffrey Lawrence, in chapter 6, examines the global processes impinging on rural Australia and altering the ways rural space is configured. This chapter traces the changes to 'family farm' agriculture in Australia, focusing upon the mechanisms – production contracts, finance capital and food industry demands – that are driving farming towards a more corporate-linked future. It also looks beyond the farm to understand the processes that are altering the structure and activities of regional communities. In Australia, the federal government's implementation of neoliberal policies has created the conditions for strong economic advancement of some spaces (notably, those close to the coast and with aesthetic appeal), but at the same time it has exacerbated the decline of many inland areas. The chapter concludes that the current forms of change in agriculture and in rural regions are leading to the depletion of both natural and social capital. Neoliberal policy settings appear to be eroding the ability of regional Australia to move to more sustainable forms of development.

In chapter 7, Richard Stayner shows how economic factors help to explain many of the observed changes in the spatial distribution and condition of rural communities. Economic changes within the farm sector and its associated industries have altered the distribution of economic activity among rural communities, as have changes within other industries represented in rural communities. The several categories of capital are examined in turn to reveal the impacts of changes in the capitals on the economic condition of rural communities. The conclusions assess the role of economic policy interventions and economic factors in the possible future of rural communities. In particular, Stayner questions whether most inland rural regions have the 'critical mass of related firms and supporting infrastructure' to provide the basis for industry clusters, while primary industries are unlikely to provide a sustainable basis for rural communities. However, fortunately for these regions, they have become 'the setting for the expression of a much wider range of aspirations and values' than those based on the extractive or productive use of natural capital, such as stewardship of the environment and appreciation of environmental and cultural amenity.

Chapters 8 and 9, by Margaret Alston, consider the social dimension of sustainability. They also address the issue of the 'other' of rural communities – the more marginalised groups, including young people, Aboriginal people, and the less economically affluent. As discussed above, the intragenerational dimensions of development and change, and particularly improvements in equity, are fundamental to the pursuit of sustainability. As Alston shows, this objective is challenged by the present reality of life for many rural residents.

Chapter 8 considers the gendering of rural community life, arguing that men and women experience rural life differently. The concept of gender is examined in an Australian rural small-town context by analysing the way that power, work, property ownership and carework are divided. In illustrating this theoretical work, the chapter draws on research undertaken in the 1990s and early 2000s and uses the five capitals framework – economic, institutional, human, social and environmental – to show that living in small Australian rural communities in the early 21st century is very much a gendered experience.

Chapter 9 examines the notion of social exclusion and its applicability to inland rural Australia. Social exclusion relates to the constraints beyond the control of an individual that restrict their ability to fully participate in society. It is argued that growing numbers of rural Australians are particularly vulnerable to social exclusion because of factors such as globalisation, neoliberal policy responses and government expectations that individuals and communities are responsible for their own futures and must become self-reliant. Growing levels of social exclusion are evident in indicators such as education access, health and employment opportunities. The lack of services and opportunities makes it particularly difficult for many rural Australians to respond effectively to changing circumstances and has led to social exclusion for many.

A central argument of chapter 10, by Chris Cocklin, is that the relationship between natural capital and primary production is a very basic one, but that the specific contours of this relationship are influenced by a wide range of factors – markets, environmental variability and change, community expectations, governance, land use change, technology, and demographic change. If we fail to understand the host of 'intervening variables', then attempts to forge a more complementary relationship between production and the environment will only succeed by luck, not design. The final section provides some consideration of the relationships between natural capital and social capital, and how these two can be brought together in support of sustainability in the context of rural communities.

PART III RESPONDING TO THE CHALLENGES

Matthew Tonts, in chapter 11, sets out to examine the role of government policy in promoting a sustainable rural Australia. He argues that the concept of sustainability has only recently been adopted by governments in relation to rural areas, and represents an important shift in thinking. Rather than focusing narrowly on issues such as economic development and productivity, sustainability provides an opportunity for governments to look at links between economic, social and ecological systems when formulating and implementing policy. A key element of sustainability is a consideration of intergenerational issues, particularly in relation to the availability of economic, social and ecological resources/services. Thus, a central issue facing government is not only how to tackle problems in an integrated way (taking into account the links between economic, social and ecological systems), but also in a way that ensures a degree of equity across generations. One of the underlying themes of this chapter is the role that government plays in developing or degrading natural, economic, social, human and institutional capital in rural areas.

Jacqui Dibden and Lynda Cheshire, in chapter 12, examine one of the roles adopted by Australian governments in recent years. Over the last decade, governments at all levels have turned their attention to encouraging programs of rural community development to deal with many of the economic, social and environmental dimensions of rural decline. As the name suggests, this approach makes 'community', as opposed to simply economic, development a prime objective. However, this new form of community development bears little resemblance to the radical social change mandate of the earlier community development movement. This chapter examines precisely what is meant by the term 'community development' today and the various ways in which it has been understood and applied by policy-makers and practitioners. In the process, the authors bring 'to light the differing, and often competing, rationalities of economic growth and social inclusion that coexist within the rural community development approach'. A critical analysis of contemporary rural community development initiatives in Australia reveals the limitations of this approach, particularly the failure thus far to achieve an integration of social, economic and environmental objectives. The capacity of rural communities to help themselves has been promoted through programs to 'build' capacity, social capital and leadership skills (but which are often only accessible to people with the skills to apply for them). Although these programs have often provided valued resources to rural people, the ability to achieve sustainability may still be constrained by factors beyond the control of local communities, notably the presence or absence of favourable location and natural and built amenity.

Ian Gray, in chapter 13, presents a more optimistic view of future possibilities. He argues that social scientists can find themselves confronted by two opposing, but both apparently reasonable, interpretations of sustainability. One interpretation is based on the observation that history is moving us all through an inevitable process of change over which we have no control, such as the view that the process of globalisation will inevitably transform all our economies and societies. In the context of sustainability, the idea that we are headed for inevitable environmental disaster is also such a view. In contrast, some people believe that globalisation is not inevitable, or that we are making significant and effective changes to ensure that our environment will continue to support life as we know it. The future of small regional communities can also be seen from similarly opposing perspectives. From one perspective, it can be argued that the economic, social and environmental systems are all steering our communities inevitably towards their demise. From the other perspective, it would appear that people can take action to change the trajectory of decline. However, when regional communities act in the interests of their people, they face a range of challenges. This chapter analyses these challenges with one eye to objective analysis and the other eye to interpretations of those challenging circumstances by the people who confront them, the residents of small regional communities.

SUSTAINABILITY AND POLITICAL CHOICE

This book examines the intertwining themes of sustainability and change, and raises questions about how communities and the natural environment can be sustained in the face of change, and what changes must be made in order to achieve sustainability. What is clear is that sustainability cannot be conflated with the preservation of the status quo. Changes are certain to occur – not all rural towns or all farms will survive, landscapes will be transformed as degraded areas are withdrawn from production, new land uses and forms of rural settlement will take their place. This does not mean, however, that the future is predetermined or that only change of a particular kind (for example, 'globalisation') will occur. The supposed 'inevitability' of certain trends is often a rhetorical flourish used to give weight to preferred policy positions rather than describing an inescapable reality.

Governments have to make choices on behalf of their constituents. Do we fund early childhood education and family support programs or build more prisons for the casualties of educational failure and family dysfunction? Do we provide financial assistance to farmers for tree planting and land retirement or deal with the consequences of the salinisation of rivers, including (the most extreme case) the loss of Adelaide's water supply? Unfortunately, the choices actually made by governments often tend to be short-term, as responses to immediate political concerns, rather than to a long-term vision of sustainability. This does not mean that there are no alternative possibilities. These might include the reintroduction of notions of the 'public good' and the adoption of longer-term outlooks, as well as more determined efforts to achieve the genuine integration of the social, economic and environmental, which would allow rural areas to reorient towards more sustainable futures.

NOTE

1   On community, see, for example, Cohen (1989); Crow & Allan (1994); Day (1998a); Liepins (2000a); (Liepins 2000b); Scott et al. (2000). On sustainability, see, for example, Barbier (1987); Cocklin (1995); Pierce (1992); Robinson (2002).

PART I

SUSTAINABILITY IN CONTEXT

2

RURAL COMMUNITIES AND SUSTAINABILITY

Alan Black

Because the terms 'community', 'rural' and 'sustainability' are each contestable, this chapter begins with an exploration of the scope of these terms. It seeks to identify key characteristics of sustainability, while recognising that there may be differences of opinion as to the balance between its different elements or dimensions. The distinction between 'weak' and 'strong' versions of sustainability is examined. So too is the concept of sustainable development. Finally, the chapter examines factors that may assist or impede the achievement of sustainable livelihoods in rural Australia.

WHAT IS A COMMUNITY?

The term 'community' has been the subject of much debate. Usage of the term can vary greatly from one context to another. In a classic article on the subject, Hillery (1955) identified 94 definitions of community and found many inconsistencies and differences of emphasis between them. Nevertheless, most of the definitions referred to a collectivity of people engaged in social interaction within a geographic area and having goals or norms in common. Although there may be debate about the boundaries of a community, about the extent to which interaction occurs among all its members, and about the degree to which they have goals or norms in common, it is generally assumed that these characteristics are present in some degree in most *communities of location* in rural Australia.

More recently, the term 'community' has also been applied to categories of people who engage in a particular purpose, task or function together, or who have some form of identity in common, though not necessarily associated with the same locality. The shared function may be related to work, education, sport or entertainment, for example. The shared identity might be that of ethnic origin, occupation, disability, age, gender, sexual orientation, religion or some other characteristic. In so far as the members of groupings such as these think of themselves as forming a community, they may be termed *communities of interest*.

This book is primarily concerned with communities of location in rural Australia. These communities may sometimes, but not always, have fairly well defined physical boundaries. In many rural areas, there will be a small, dense area of housing, surrounded by areas in which housing is less dense. Occupants of these dwellings, whether in the township or beyond, may use the same sets of services, be subject to the same local government authority, and identify with much the same community. Nevertheless, for some purposes, members of that community may be linked to larger communities nearby. For example, people may do some of their shopping in the small centre, but travel to a larger centre for major items. Children may undertake their primary schooling in the small centre, but their secondary education in the larger centre. In other words, people may simultaneously be part of several communities defined at different geographical scales.

Unless otherwise indicated, when the term 'community' is used in this chapter it is assumed that:

- a community is a social construct incorporating four elements: people, meanings, practices and spatial configuration (Liepins, 2000a, 2000b)

- there may be varying degrees of agreement or disagreement among members of a community as to what such membership implies about their identities, connections, beliefs, values and practices

- the practices through which such meanings and interpretations are developed and expressed may include both co-operation and contestation, whether in formal or informal settings

- community boundaries may be well or ill defined, and people may think of themselves as belonging to – or excluded from – a variety of communities of different types or of different geographic scales (Cohen, 1985; Jones, 1995)

WHAT AND WHERE IS RURAL AUSTRALIA?

It is difficult to achieve consensus on the meaning of the term 'rural'. Definitions have variously been based on:

- *socio-demographic characteristics*, such as population density or settlement size
- predominant forms of *economic activity or land use*, such as agricultural, pastoral and other primary industries
- *socio-cultural characteristics*, such as particular kinds of social relationships and values

Each of these types of definition has attracted criticism. Definitions of the first type have been criticised for their arbitrariness: where does one set the dividing line between the rural and the non-rural, and is this point of division appropriate for all purposes? Definitions of the second type have been criticised on the grounds that a substantial and increasing proportion of the population in areas that we would usually call rural is not directly or exclusively engaged in primary production, and there is often a diversity of land uses. Definitions of the third type have been criticised as painting a romanticised or inaccurate picture of the differences between rural and non-rural (for example, urban) sections of society (see Hoggart & Buller, 1987; Pahl, 1966).

Because of the different definitions adopted in member countries, the OECD has proposed a working definition that would facilitate comparisons between countries (Dax, 1996). Its primary criterion is population density: rural areas have no more than 150 inhabitants per square kilometre, except in Japan, where the upper limit is 500 inhabitants per square kilometre. Once communities have been categorised into *rural* and *urban* on the basis of population density, regions that might include both rural and urban communities are classified into three types:

- *predominantly rural regions*, where more than 50 per cent of the population lives in rural communities

- *significantly rural regions*, where the proportion of the population in rural communities is 15–50 per cent

- *predominantly urbanised regions*, where less than 15 per cent of the population is in rural communities

Although the Australian Bureau of Statistics, together with the former Department of Primary Industries and Energy, has done some exploratory work in applying these criteria to Australia, this work has had little impact on domestic policy development.

In recent decades, the Australian Bureau of Statistics (2002a) has generally used a four-fold system to classify settlement types:

- *major urban areas*: urban areas with a population of 100 000 and over
- *other urban areas*: urban areas with a population of 1000 to 99 999
- *bounded rural localities*: population clusters of 200 to 999 people in a minimum of 40 occupied non-farm dwellings with a discernible urban street pattern
- *rural balance*: areas not included in the previous three categories

For some purposes, the first two categories are aggregated into a single *urban* category. Similarly, the third and fourth are sometimes combined into a single *rural* category. This is a much narrower definition of the rural than that adopted by the Department of Primary Industries and Energy and the Department of Human Services and Health (1994) in their jointly developed Rural, Remote and Metropolitan Areas (RRMA) system of classification. This system of classification defined metropolitan areas as urban centres with a population of at least 100 000. All other areas of Australia were classified as rural unless they met specified criteria of remoteness. The index of remoteness took account both of the average distance between residents and the distance of the statistical local area (SLA) from an urban centre offering a wide range of goods and services. Thus *rural* areas were defined as non-metropolitan areas that did not meet the criteria of being *remote* areas. This system of classification had two implications: (a) urban areas with populations up to 99 999 could be classified as rural, but (b) a particular SLA could not be classified as both rural *and* remote. These two implications are somewhat at odds with everyday use of the term 'rural'. For discussion of other limitations of this system of classification, see Bamford and Dunne (1999).

In *Prospects and Policies for Rural Australia*, Sorensen and Epps (1993: 2) defined rural as comprising all parts of the nation except the capital cities, the Gold Coast, the New South Wales Central Coast, Newcastle, Wollongong and Geelong. In its *Blueprint For Rural Development*, the National Rural Health Alliance (1998) adopted a fairly similar definition, but with Townsville also excluded. Under this definition, rural Australia is very diverse, including not only farming areas but also agricultural service centres, mining towns, coastal communities attracting holiday-makers and retirees, Aboriginal outstations, remote islands, wilderness and desert areas, and many of the major regional centres.

Major regional centres, such as Townsville, Cairns, Toowoomba and Albury-Wodonga, as well as other cities and towns with a population of 10 000 or more, are not a primary focus of this book. The focus is, rather, on smaller towns (especially those with less than 5000 residents) and rural areas as defined by the Australian Bureau of Statistics. As the wellbeing of people in these communities is linked in various ways with what is happening elsewhere, attention will also be given to these wider linkages.

WHAT IS SUSTAINABILITY?

In everyday language, the term 'sustain' may include one or more of the following connotations: support, keep in existence, supply with the necessities of life, and continue without lessening. In recent years, the adjective 'sustainable' or its opposite 'unsustainable' has been applied to a wide variety of phenomena, such as social practices, farming systems, resource use, economies, communities and societies. Sustainability can be defined broadly as the 'ability to continue an activity or maintain a certain condition indefinitely' (Eckersley, 1998: 6). So, for example, the World Conservation Union (IUCN), the United Nations Environment Program (UNEP) and the World Wide Fund for Nature (WWF) (1991: 211) define sustainable use as 'use of an organism, ecosystem or other renewable resource at a rate within its capacity for renewal'.

Meadows et al. (1992: 209) define a sustainable society as 'one that can persist over generations, one that is far-seeing enough, flexible enough and wise enough not to undermine either its physical or its social systems of support'. The word 'flexible' in this definition clearly implies that a sustainable society is not necessarily static. A society or community may change in order to respond to various contingencies. For instance, it may change because its resource base has been damaged or depleted; in which case, one may ask whether the previous organisational forms and activities would have been sustainable in the long term. On the other hand, a society or community may change because new resources have been identified or new technologies developed. Here, too, one could ask whether the new organisational forms and activities are sustainable in the long term.

When applied to communities or societies, sustainability is increasingly being seen as involving three pillars or dimensions – the economic, the social and the ecological. The *ecological dimension of sustainability* has to do with the extent to which ecological systems – on which all life depends – are capable of continuing to perform their essential functions into the future. This dimension of sustainability relates, for example, to the extent to which:

- ecosystem integrity is preserved
- biological diversity is maintained
- rates of use of renewable resources do not exceed regeneration rates
- rates of waste generation or pollution emission do not exceed the assimilative capacities of the environment

The *economic dimension of sustainability* has to do with the extent to which economic systems are capable of continuing for the long term. Examples of this dimension of sustainability are the degree to which:

- systems of production, exchange and consumption can continue
- satisfactory standards of living for all are being achieved now and can be maintained
- rates of use of non-renewable resources do not exceed the rate at which sustainable renewable substitutes are developed
- economic systems are able to adapt to various contingencies, such as fluctuating environmental conditions (for example, rainfall, temperature, geothermal activity), demographic changes and technological developments

The *social dimension of sustainability* has to do with the extent to which social values, social identities, social relationships and social institutions are capable of being maintained into the future. This dimension of sustainability can be illustrated by the extent to which:

- there are some widely accepted and enduring norms or values, such as reciprocity, procedural equity and respect for law
- both individual identity and cultural diversity can be maintained
- social institutions are able to make a continuing contribution to the fulfilment of people's needs
- social institutions are able to adapt to various contingencies, such as fluctuating environmental conditions, economic changes and technological developments

These three dimensions of sustainability are seen as interrelated. Social and economic systems generally have an impact upon ecological systems, and vice versa. There is also a two-way interaction between economic systems and social systems. Appropriate integration between the three dimensions is required in the quest for sustainability (Dale & Hill, 2001; Giddings et al., 2002).

SUSTAINABLE COMMUNITIES

What, then, are the essential characteristics of a sustainable community? Here is a description proposed by Richardson (1994):

> A sustainable community has a stable, dependable and diversified economic base that does not over-stress the carrying capacity of natural systems, maintains the supply and quality of renewable resources, and strives continually to reduce its demands on non-renewable resources. Its economy provides both a range of opportunities for rewarding work, and a level of prosperity on the basis of which, equitably shared, the community actively and continuously works to satisfy the basic needs of every one of its members and to provide each with the opportunity to fulfil his or her potential, within a supportive social environment, a safe, liveable physical environment, and a clean, healthy, vital natural environment. A sustainable community does not achieve or maintain its own sustainability at the cost of the sustainability of other communities/ecosystems, including that of the broader community/ecosystem of which it is a part.

This description, which integrates ecological, economic and social dimensions, could be seen as an ideal towards which a community might strive. There could, of course, be debate about the extent to which some of the listed characteristics are essential for community sustainability. For example, how essential is a diversified economic base? Economic diversity, though perhaps not essential, is nevertheless likely to give a community greater capacity to withstand changing circumstances in the wider economy, thus enhancing its prospects of being sustainable in the long term. Economic diversity can also help a community to be more self-sufficient, rather than being heavily dependent on goods and services produced elsewhere. Such self-sufficiency may be ecologically beneficial at global and national levels if it results in a global reduction of energy consumption (for example, for transportation) and a consequent lessening of pollution emission.

The Sustainable Community Roundtable (1999), established in South Puget Sound (USA), proposed the following vision of what a sustainable community would be:

> A sustainable community continues to thrive from generation to generation because it has:
> - a healthy and diverse *ecological system* that continually performs life sustaining functions and provides other resources for humans and all other species
> - a *social foundation* that provides for the health of all community members, respects cultural diversity, is equitable in all its actions, and considers the needs of future generations
> - a healthy and diverse *economy* that adapts to change, provides long-term security to residents, and recognizes social and ecological limits.

This description again brings together ecological, social and economic dimensions of sustainability. As a vision it contains much that is commendable. In a small survey conducted in South Puget Sound in 1999, it was found that most respondents generally accepted the Roundtable's vision. There were, nevertheless, some constructive suggestions as to how the vision could be improved, and some people questioned the necessity of including one or another feature of it (Beck, 1999). This illustrates the difficulty of attaining complete consensus on what is meant by the term 'sustainable community' and what are the conditions needed to attain it.

A briefer description is contained in a publication of the New South Wales Premier's Department Strengthening Communities Unit (2001: 28):

> Sustainable communities ... maintain and improve their social, economic and environmental characteristics so that residents can continue to lead healthy, productive and enjoyable lives. Sustainable development in these communities is based on an understanding that a healthy environment and a healthy economy are both necessary for a healthy society.

This description, like the two previous ones, refers to the interdependence of economy, society and environment. It, too, uses the notion of health as part of its description. While most people would agree with the desirability of the general ideals contained in the description, differences of opinion may exist over the criteria to be used in assessing what is 'a healthy society', 'a healthy economy' and 'a healthy environment', and especially as to what compromises, if any, may be made between these objectives. Moreover, different environments, societies and economies are apparent at different spatial scales. There are complex interactions between phenomena at these different levels of aggregation (Cocklin et al., 1997; Giddings et al., 2002).

To summarise this section: sustainability is an example of what Gallie (1956) termed 'a contested concept'. This is so because people may give competing answers to questions such as: What is to be sustained? For whom? Over what scale of time and place? How? Why? Even when people agree on the answers to many of these questions, it may not be possible simultaneously to maximise each dimension of sustainability. In that case, opinions may differ on the relative priority to be accorded to different dimensions and on the extent to which trade-offs may be made between them.

'WEAK' AND 'STRONG' SUSTAINABILITY

In the debate about forms of, and conditions required for, sustainability, a key issue has been the question of whether a high degree of substitutability is possible between natural resources on the one hand and (humanly) produced capital on the other. Daly and Cobb (1989: 72), building on earlier discussions of this question, have drawn a distinction between what they termed *weak sustainability* and *strong sustainability*. 'Weak sustainability' is the variant represented in the writings of neoclassical economists such as Solow (1974, 1986), Hartwick (1977, 1978a, 1978b, 1990), Dixit et al. (1980), Becker (1982), Dasgupta and Mitra (1983) and Withagen (1996), who have examined the conditions under which a society's economic wellbeing can be maintained despite declining stocks of non-renewable resources. One result of this investigation is what has come to be known as Hartwick's Rule, namely that economic wellbeing can be maintained constant over time if the current economic returns from the use of flows of non-renewable resources are invested in reproducible assets such as machines. Hartwick's Rule is based on assumptions of constant population, a closed economy[1] and sufficient substitutability between natural resources (exhaustible stocks) and produced capital (reproducible assets).

Consistent with that line of reasoning, Solow (1993) contends that the quest for sustainability is concerned, in essence, with ensuring the continuation of an economy's productive capacity so as to allow each future generation the option of being as well-off as the generation that preceded it. As Solow (1986: 142) argued in an earlier paper:

> The current generation does not especially owe its successors a share of this or that particular resource. If it owes anything, it owes generalized productive capacity or, even more generally, access to a certain standard of living or level of consumption. Whether productive capacity should be transmitted across generations in the form of mineral deposits or capital equipment or technological knowledge is more a matter of efficiency than of equity.

In Solow's opinion, natural resources can safely be used up or run down, provided that sufficient investment is made in alternative forms of income-generating capacity. This version of sustainability is termed 'weak sustainability' by Daly and Cobb (1989).

By contrast, Daly and Cobb (1989) advocate 'strong sustainability'. In their view, the resources and services provided by nature (sometimes spoken of as 'natural capital') on the one hand and humanly produced capital on the other are *complements* rather than *substitutes* in most production functions; the concept of strong sustainability requires that each of these forms of capital be 'maintained intact'.

The requirement that natural capital be 'maintained intact' means, according to Daly (1994: 24), that in some aggregate sense natural capital should remain constant. This principle leads Daly (1995: 50) to propound several rules for the management of production and consumption. With some rephrasing, these rules are as follows:

Output rule:
Waste outputs should not exceed the natural absorptive capacities of the environment (that is, nature's *sink* services should not be depleted).

Input rules:
(a) For renewable inputs, harvest rates should not exceed regeneration rates (that is, nature's *source* services should not be depleted).

(b) For non-renewable inputs, the rate of depletion should not exceed the rate at which renewable substitutes can be developed.

(c) If a renewable stock is consciously divested (i.e., exploited non-renewably), it should be subject to the rule for non-renewables.

Although rule (b) does envisage some substitutability, Daly insists that this should be a real substitution, not a merely financial one. For example, the income from depletion of fossil fuels should be invested in the development of new energy supplies from renewable sources.

Daly (1995) rejects the suggestion that the strong version of sustainability implies that no species should ever be allowed to become extinct or that any non-renewable source should ever be used. To insist on these additional conditions would, in his view, be to advocate 'absurdly strong sustainability'. Nevertheless, Holland (1997) has argued that no distinction of any substance can be drawn between weak and strong sustainability as defined by Daly, but that a version of what Daly has dubbed 'absurdly strong sustainability' is not necessarily absurd after all.

Controversy over the distinction between 'weak' and 'strong' versions of sustainability has resulted in numerous publications, including several papers in two recent collections (Dobson, 1999; O'Neill et al., 2001) plus a book devoted wholly to the topic (Neumayer, 1999). Proponents of the so-called weak version of sustainability do not concede that their perspective is morally or intellectually weak or that it is inferior to Daly's purportedly strong version. On the contrary, they see their own position as more defensible than his.

I would argue that proponents of 'strong sustainability' are correct when they state that there are physical limits to various natural resources and to the stability and resilience of ecosystems. Likewise, there are limits to humans' capacity to find or develop substitutes for these resources or for the services provided by existing ecosystems. But it is often difficult to know precisely what those limits are. The fact that humans have found or developed substitutes for some resources in the past is no guarantee that humankind will always be able to do so in the future. Conversely, a present inability to find or devise a substitute does not necessarily imply that no such substitute will ever be found or developed. The weak-versus-strong debate indicates again that sustainability continues to be a contested concept.

SUSTAINABLE DEVELOPMENT

Although the term 'sustainable development' has been widely used since the late 1980s, there have been many differences of opinion as to what it implies. An influential definition and exposition of the concept was given in *Our Common Future*, the report of the World Commission on Environment and Development (1987) – often referred to as the Brundtland Report. That report defined sustainable development (p. 43) as 'development that meets the needs of the present without compromising the ability of future generations to meet their own needs'. In elaborating on this concept, the report stated (p. 46) that sustainable development is:

> ... a process of change in which exploitation of resources, the direction of investments, the orientation of technological development, and institutional change are all in harmony and enhance both current and future potential to meet human needs and aspirations.

A major focus of *Our Common Future* was on ways of better fulfilling the needs of the world's poor while also protecting the environment. The report argued that both rich and poor countries should seek sustainable pathways for development that would achieve imperatives such as the following: reviving economic growth; changing the quality of growth to make it less material- and energy-intensive and more equitable in its impact; meeting essential human needs for means of livelihood, food, water, clothing, energy, shelter, sanitation and health care; ensuring a sustainable level of population; conserving and enhancing the resource base; reorienting technology and managing risk so as to pay greater attention to environmental factors; and integrating economic and ecological considerations in decision-making.

The publication of *Our Common Future* was one of the influences that led the Commonwealth of Australia to issue a discussion paper titled *Ecologically Sustainable Development* and to develop a *National Strategy for Ecologically Sustainable Development* (Commonwealth of Australia, 1990, 1992).[2] The discussion paper stated that:

> Ecologically sustainable development means using, conserving and enhancing the community's resources so that ecological processes, on which life depends, are maintained, and the total quality of life, now and in the future, can be increased. (Commonwealth of Australia, 1990: iii)

The core objectives identified in the *National Strategy for Ecologically Sustainable Development* (Commonwealth of Australia, 1992: 8) were:

- to enhance individual and community well-being and welfare by following a path of economic development that safeguards the welfare of future generations

- to provide for equity within and between generations
- to protect biological diversity and maintain essential ecological processes and life-support systems

To achieve these objectives, the Strategy (p. 8) adopted seven guiding principles. These focused on integrating both long-term and short-term economic, environmental, social and equity considerations in decision-making; adopting a precautionary approach where there is a possibility of serious or irreversible environmental damage; taking account of global environmental impacts of actions and policies; developing a strong, growing and diversified economy that can enhance the capacity for environmental protection; maintaining and enhancing international competitiveness in an environmentally sound manner; using cost-effective and flexible policy instruments, such as improved valuation, pricing and incentive mechanisms; and providing for broad community involvement in these issues.

In addition to listing a range of specific policies for implementation by Commonwealth, state and territory governments, the *National Strategy for Ecologically Sustainable Development* incorporated as Appendix B the 27 principles that had been adopted by the United Nations Conference on Environment and Development in 1992. Appendix C of the *National Strategy* gave a brief outline of another document, titled *Agenda 21*, that had been endorsed at the same United Nations conference. *Agenda 21* was a detailed blueprint for action to achieve sustainable development, which it saw as involving three 'interdependent and mutually reinforcing pillars': economic development, social development and environmental protection. It called for all tiers of government and all segments of society to help achieve sustainable development. Chapter 28 of *Agenda 21* dealt specifically with the roles of local government and local communities. It proposed that each local authority should enter into dialogue with its citizens, community organisations and private enterprises in order to achieve consensus on a plan of action – a 'local Agenda 21' (LA21) – for their community (United Nations Conference on Environment and Development, 1992).

Despite the hope expressed in *Agenda 21* that most local authorities would have developed LA21 action plans by 1996, less than 10 per cent of Australian local government authorities (LGAs) succeeded in doing so by the target date (Addison, 2001). A manual to help LGAs develop such action plans was prepared by Environs Australia (the Local

Government Environmental Network) and published in 1999 by the Commonwealth government, with a foreword by the federal Minister for the Environment and Heritage (Environs Australia, 1999). While more LGAs have formally adopted LA21 plans since then, the vast majority have not. Very few LGAs in rural Australia have explicitly done so, although most have adopted policies on some aspects of environmental protection. Some have also undertaken initiatives to foster economic and social development in their locality. There is further discussion of such initiatives in chapters 11 and 12.

SUSTAINABLE LIVELIHOODS

The World Commission on Environment and Development (1987: 57) and the United Nations Conference on Environment and Development (1992) underlined the need for sustainable means of livelihood. With reference primarily to Third World countries, writers such as Chambers and Conway (1992), Carney (1998) and Scoones (1998) have spoken of *sustainable rural livelihoods*. The Institute of Development Studies has adopted the following definitions (Scoones, 1998: 5):

> A livelihood comprises the capabilities, assets (including both material and social resources) and activities required for a means of living. A livelihood is sustainable when it can cope with and recover from stresses and shocks, maintain or enhance its capabilities and assets, while not undermining the natural resource base.

Like other concepts of sustainability, the notion of sustainable livelihoods is made up of multiple, and sometimes contested, elements (Scoones, 1998; Sneddon, 2000). Nevertheless, it points to some issues pertinent to an understanding of the sustainability of rural communities in Australia and elsewhere. Substantial numbers of farm households in Australia are partly dependent on off-farm income to survive. The proportion of farmers and spouses who work off-farm has doubled since the 1980s, with 34 per cent of farmers or spouses doing so in 1995/96 (Garnett & Lewis, 1999: 3). In broadacre farming in 1996/97, men spent an average of 48 hours in on-farm work with another six hours in off-farm employment, while the figures for women were 16 and eight hours respectively (Garnett & Lewis, 1999: 17). The majority of women with off-farm employment (especially those with post-school qualifications) occupied managerial, professional or semi-professional positions, whereas the largest grouping of men did labouring jobs, mainly for other farms. More than half the farm women with off-farm employment worked in the education, health or community sectors (Garnaut et al., 1999).

The stability and adequacy of on-farm income is influenced not only by factors such as the size, fertility, location and mode of operation of the farm but also by the national and international supply of, and demand for, the commodities produced. Questions about the sustainability of livelihoods in the farming sector tend to gain prominence in public discussion when catastrophic events ('shocks') such as severe droughts, unexpected slumps in commodity prices or foreclosures on bank loans force farmers from their properties. But long-term sustainability is also affected by various 'stresses' – gradually occurring phenomena that may nevertheless have a large cumulative effect – such as declining soil fertility, increasing salinity, global warming, or evolving consumer preferences.

Such shocks and stresses may also have direct or indirect impacts on the sustainability of livelihoods in rural towns. For example, in the case of small inland towns whose economy is heavily dependent on servicing agriculture, reductions in the prosperity of farmers may trigger or aggravate a vicious circle of decline in which there is business contraction, out-migration, a withdrawal or downgrading of various services, the erosion of local employment opportunities, and further out-migration (Sorensen, 1993). Out-migration will not necessarily be to metropolitan areas. Some may flow to larger regional centres (so-called 'sponge' cities) or to coastal areas found to be attractive for one reason or another.

The sustainability of livelihoods in rural towns is not dependent solely on the prosperity of the farming population. In varying degrees, the economy of these towns has been decoupled from agriculture (Stayner & Reeve, 1990). Diverse factors, some local and some much wider, are likely to affect the livelihood resources available and the livelihood strategies adopted in these towns. For example, some small inland towns are heavily dependent on timber milling or mining. The sustainability of livelihoods in these towns is subject not only to market trends for the commodities produced but also to the longevity of the resource base, and perhaps to specific government policies such as Regional Forest Agreements. Where ownership of mines or mills is distant from the local community, this can be an additional complication. Some small towns have endeavoured to diversify their economies so as to offer a wider range of employment opportunities for their people, thus being less subject to economic downturns in a single industry.

The welfare system in Australia is designed to provide some financial support to people who lack adequate means of livelihood. Although there have been some fluctuations from time to time due to the prevailing economic conditions, the proportion of the population aged at least 16 years and reliant on welfare benefits (aged pension, unemployment benefit, disability pension or sole parent benefit) tended to increase during the 1970s and early 1980s, and again during at least the first half of the 1990s. The proportion receiving one or other of those four welfare benefits rose from 12.5 per cent in 1973 to 23.1 per cent in 1993 (ABS, 1994: 147). Such people make up a higher percentage of the adult population in non-metropolitan regions than in metropolitan areas of Australia. Hugo and Bell (1998) argue that this difference is due to factors such as the generally lower cost of living (especially for housing) in non-metropolitan regions, as well as the lifestyle attractions of non-metropolitan coastal locations. Over the period from 1986 to 1991, for example, there was a net *inflow* of low income (<$16001) households to the Central West, Murray/Murrumbidgee and Far West/North Western regions of New South Wales, as well as to the Hunter, South Eastern and North Coast regions of that state. This contrasts with a net *outflow* of high income (>$50000) households from five of those six regions (Hugo & Bell, 1998: 125).

Consistent with these data, unemployment rates since the mid-1980s have generally been higher in non-metropolitan regions than in metropolitan areas (ABS, 2001: 140–41). A more fine-grained analysis reveals that there are considerable regional differences in labour markets and unemployment. In some small inland towns, especially in the wheat–sheep belt and in some mining areas, higher than average rates of unemployment have been associated with a decline in the total number of jobs available locally. By contrast, some coastal parts of non-metropolitan New South Wales and Queensland have simultaneously experienced relatively high rates of growth in the number of persons employed, and relatively high rates of unemployment (NIEIR, 1998). As well as the lifestyle reasons that make such locations attractive, it appears that there is a tendency for in-migration to overshoot the number of jobs available.

Movement of people into or out of a community may indicate various trends in 'people prosperity' and 'place prosperity' (Bolton, 1992). The longer the period of analysis and the greater the movement of people, the more important the distinction between 'people prosperity' and 'place prosperity'. By definition, upwardly mobile people are improving in prosperity. If such people are leaving a community, and if there is a net loss of positions there at the level they had occupied, the improved prosperity of the upwardly mobile individuals may, in some cases, be accompanied by a decline in the overall prosperity of the community they have left. This situation has occurred in some small country towns as a result of bank, school or hospital closures and the concentration of administrative, commercial and other functions into larger regional centres. Of course, these agglomerative tendencies have not necessarily resulted in upward or even horizontal mobility for all people who had been providing those services in small towns. Some of these people have simply been declared redundant.

For state and federal governments there is an ongoing issue of the relative emphasis to be placed on 'people prosperity' and 'place prosperity' in the design of public policy. For example, to what extent, and in what ways, should governments try to arrest or reverse the decline of small towns? Should the processes of decline or growth simply be left to market forces? If governments have a role to play in facilitating processes of adjustment, should the main emphasis be upon 'bringing jobs to people' or 'bringing people to jobs'? Where do issues of equity and access come into all this? For discussions of some of these issues, see Sorensen (1993), Collits (2001), Forth (2001), Yenken and Porter (2001) and Tonts, in chapter 11.

CONCLUSION

The sustainability of rural communities is problematic for several reasons. First, there is the matter of definition. What and where is the rural? What constitutes a community? What is sustainability? To each of these questions people have given a variety of answers, some of which have been examined in this chapter. Particular attention has been given to the integration of economic, social and ecological dimensions of sustainability, to the contrast between 'strong' and 'weak' criteria for sustainability, and to the concepts of sustainable development and sustainable livelihoods. This provides a backdrop to the more detailed examination of various aspects of rural sustainability in later chapters.

A second set of reasons why the sustainability of rural communities is problematic is because trends in the wider economy, society and environment have had adverse or mixed effects on many rural communities, especially inland settlements with fewer than 2000 inhabitants. These trends include demographic changes, such as lower fertility, ageing populations and changing household structures; technological developments affecting modes of production, transportation and communication; social and cultural changes in consumption patterns, social relationships and values; economic trends, such as deregulation, privatisation and globalisation; the consolidation of educational, health, financial and other services into larger regional centres; and environmental trends, such as the spread of dryland salinity, the over-burdening of water resources and the loss of biodiversity. The impacts of changes such as these will also be explored in more detail in subsequent chapters.

NOTES

1   Asheim (1986), Hartwick (1995) and Sefton & Weale (1996) have identified amendments to the Rule for open economies trading with one another (Neumayer, 1999).
2   For accounts and assessments of the development of that Strategy, see Beder (1996) and Diesendorf & Hamilton (1997b).

3

RURAL SUSTAINABILITY IN HISTORICAL PERSPECTIVE

Graeme Davison

'Sustainability' is the newest addition to the lexicon of rural decline. For more than two centuries, Australians have wrestled with a deep dilemma. Almost from the beginnings of European settlement, the coastal cities claimed a large and apparently abnormal proportion of the population of the colonies. Yet the countryside remained, in the eyes of many observers, the only true and enduring source of national prosperity and wellbeing. Settlers drawn from greener, wetter European lands, with strong traditions of village and small-town life, were ill at ease in a dry, brown land where geography seemed to dictate sparser and more fragile patterns of settlement. They could never quite throw off the hopeful assumption that, while their rural communities were for the moment small and dispersed, they would some day assume the 'natural' European pattern.

For most of the last century, however, the cards have been stacked against the towns of inland Australia. The 'Bush', long a national ideal, has become in our times an object of pity. Sustainability, with its assumptions of maintenance rather than growth, and of an environmental, rather than economic or political, frame of reference, represents a belated climb-down from the more sanguine expectations of earlier generations. Tracing the history of that shift enables us to recognise more clearly both the historical roots, and the nature, of our current dilemma.

Recent studies of rural discontent acknowledge, but seldom pursue, the historical dimensions of current problems.[1] Their framework of reference is often short, just a decade or so; yet the problems themselves are often long-term and intractable. In this respect, current research may be out of step with the outlook of rural communities themselves, which often have long memories, kept alive by the all-too-tangible reminders of better times evident in disused or decaying buildings and faltering local institutions. In the 1940s, one of the wisest students of Australian rural society, Professor Samuel Wadham, observed that many rural communities were already 'historic relics'. 'The first point in making a survey of a community consists in getting the correct perspective of its history' (Wadham, 1943a: 64–65).[2] What was then true of individual communities remains true of communities in general. Only by excavating the hopes of the past can we begin to appraise the depth of community unease and the prospects for sustainability. For sustainability is not only an economic, demographic and environmental concept; it is also a historical concept. It links future expectations and present policies with past experience.

Over the past two centuries, thinking about Australian rural communities has undergone a series of distinct phases, which I summarise here and expand in the rest of the chapter. In the opening decades of British settlement, colonial administrators *planted* new towns, confident that they would naturally evolve towards European maturity. In the later 19th century, the golden age of liberalism, colonial governments sought to *water* the soil in which rural communities could grow, by opening opportunities for closer settlement ('unlocking the lands') and supplying the infrastructure of railways, irrigation and education that would enable individual and collective enterprise to flourish. By the early 20th century, however, young people were drifting from the countryside to the city, and the inland towns had begun their long decline. Progressive reformers, mindful of the strong contribution of rural enterprise to national prosperity and fearful of the corrupting influences of city life on the 'future of the race', now sought to *protect* rural communities through a range of special government programs. These included economic supports, such as tariff protection and preferential rail freight rates, as well as social programs in education, health, broadcasting and rural services (see also Tonts, in chapter 11). The Second World War and the dawn of the atomic age reinforced fears that the nation was endangered by excessive concentration of population in the coastal cities. In the heyday of Postwar Reconstruction, there was strong, if short-lived, confidence that the state could *plan* and support rural communities through comprehensive programs of decentralisation. Despite a brief revival of decentralist thinking during the Whitlam years, however, the coastal cities continued to expand and rural communities to slowly decline. By the 1970s, the contribution of the rural sector to national prosperity had dramatically shrunk, and, with it, the moral claim of the countryside as the fount of national virtue. With the parallel decline of the National Party, there was little, now, to protect rural Australia from the chill winds of neoliberal ortho-doxy. Rather than being supported, as they had been for so long, by state subsidy, rural communities now faced the challenge, not only to survive, but to *sustain* themselves economically and environmentally.

The concept of 'rural community' implies a commonality of interest and outlook between farmers and rural townsfolk that was often chal-lenged in the past. The rural community or small town occupied an intermediate place in what was often characterised as the great divide in Australian society between the City and the Bush. The binary logic of Country and City, with its age-old echoes of Innocence and Experience, is unable to accommodate places that lie awkwardly between the two (Williams, 1975: 71). In 1962, an American visitor observed that: 'The country township is the forgotten feature of the frontier' (Meinig, 1970: 166). Another 40 years of decline has not saved it from that neglect.

Yet, only a century ago, non-metropolitan communities of between 500 and 20 000 people claimed about one-third of the Australian popu-lation, as many as those who lived on farms or in the capital cities (McEwen, n.d.). A historian of towns in the South Australian wheat belt shrewdly observes, 'Towns grew up where farmers wanted them and grew to the extent that farmers supported them' (Jones, 1991: 98). Farmers often took a limited and strictly pragmatic view of the value of the community facilities embodied in neighbouring towns. In the 1880s, for example, the coming of the railway often precipitated a clash between the ambitions of farmers, who appreciated the opportunity to buy direct from city suppliers, and townsmen, who feared the extinction of local business. In the 1940s, a social survey of Victorian country towns detected a similar gap between the interests and outlooks of farmers and townies, and an underlying fear on the part of the towns-folk. 'Insecurity is one of the most dominant emotions in country towns', it found (McIntyre & McIntyre, 1944: 268).

PLANTING

European Australia began as an archipelago of convict settlements distributed at about 1000-kilometre intervals along a rugged coastline. The land beyond this 'fatal shore' was often forbidding and a long mountain range along the east coast impeded easy passage to the interior. Forty years after the foundation of New South Wales, more than 40 per cent of the population was located in Sydney and the next seven largest towns contained between them less than half the population of the capital itself (Butlin, 1964a: 152–56). Early settlers arrived with the expectation that eventually the new land would become more closely settled, that towns and villages would grow up and that an urban hierarchy, somewhat like the European one, would emerge, with a mixture of villages, market towns, provincial cities and coastal towns. On their tours of the interior the early governors, like Lachlan Macquarie, paused to select the sites for new towns, and to supervise the process of laying them out, usually on plans they had borrowed from other parts of the Empire, especially India.[3]

The towns they planted in the interior, however, grew only slowly, and sometimes hardly at all. European settlers were beginning to learn what the Aborigines had discovered during more than 40 000 years of occupying the continent: that its dry climate and thin soils were ill-fitted for intensive agriculture.[4] By the 1830s, it was plain that extensive grazing, supported by a small and largely itinerant labour force, was likely to be the staple form of economic activity (McCarty, 1964: 1–24). The hopes of some early governors, like King and Bligh, to create a prosperous peasantry in New South Wales were soon blighted (Fletcher, 1969: 191–218).

The early colonists' views of town and country life were strongly influenced by their often contradictory assumptions about the effects of the concentration and dispersion of population. Concentration could denote both healthy sociability and harmful contagion. Dispersion could signify both benign liberty and dangerous licence. Most observers believed that decentralisation was the natural form of settlement for a free society. The independent landowner, exercising a kindly superintendence over a small rural community, was the ideal both of English Whigs and American Republicans like Thomas Jefferson. But if dispersion was the natural condition of freemen, concentration was often the inevitable form for a settlement of unfree men. Most Australian convicts worked as assigned servants outside the walls of gaols, but the need to keep them under surveillance and supervision meant that they could not be permitted to range too widely beyond the gaze of their guardians.

Under Governor Lachlan Macquarie, a large number of convicts were employed on public works and other occupations within the city of Sydney. By 1819, his critics, especially the emerging class of free settlers, had prompted the British government to appoint Commissioner John Thomas Bigge to investigate his administration. Their criticisms often centred on the undesirable effects of concentrating convicts in the towns. The colony's largest landowner and pioneer of the wool industry, John Macarthur, argued (Ritchie, 1971: 73–74):[5]

> When men are engaged in rural occupations their days are chiefly spent in solitude – they have much time for reflection and self examination, and they are less tempted to the perpetration of crimes than [they] would herded together in Towns, amidst a mass of disorders and vices.

Macarthur had economic as well as moral reasons for wanting to disperse the convict population to the interior. He envisaged New South Wales as a plantation economy of large pastoral estates worked by a well-disciplined convict workforce.

Optimistic observers noted the seemingly abnormal concentration of population in the coastal towns, and the relative weakness of the inland towns, but regarded it as an infantile stage of development that would eventually be outgrown. Colonial development was an evolutionary process, in which society progressed naturally from hunting and gathering, through pastoralism and agriculture to industry and urbanisation (Hamer, 1990: 91–96). But while agreeing on the overall pattern of development, they were less unanimous on the timetable and whether it was possible to accelerate it. The most systematic and influential theorist of colonial development, Edward Gibbon Wakefield, believed that the principles of 'concentration' and 'dispersion' could be brought into balance by adjusting the supplies of land and labour so as to create a society of small farms and towns spread evenly across a prosperous countryside. Writing his *Letter from Sydney* in London's Newgate prison, however, Wakefield (1929) failed to recognise the many local complications governing the process of community formation in a new country. His critics, who included the great population theorist Thomas Malthus (Select Committee on the Disposal of Lands, 1836), believed that his schemes encouraged a premature and unsustainable degree of concentration. Colonial administrators, mindful of geographical and political realities, took a more sober view of the prospects for closer settlement. New South Wales appeared to be 'marked out by nature for a Pastoral Country', Lord Glenelg, Secretary of State for the Colonies, had concluded by 1836 (Hamer, 1990: 111).

Its geography presented a 'physical impediment to the close concentration of its inhabitants with which it would be only futile to contend by human laws'. Yet, while they considered it impossible to legislate the growth of rural communities, colonial authorities often acted as though they believed that they could at least lay the foundations for future development. Thomas Mitchell, Surveyor-General of New South Wales and one of Wakefield's strongest critics, drew plans for no fewer than 60 new inland townships (Hamer, 1990: 104).

By the early 1840s, some optimistic observers believed that nature was at last taking its course. In his *Analytical View of the Census of New South Wales,* Ralph Mansfield (1841), editor of the *Sydney Morning Herald,* noted the steady growth of inland and coastal towns, which had recently outstripped the growth of Sydney. The increase of the other towns, he believed, illustrated 'that peculiar principle in the political economy of Australia, so imperfectly understood in the mother country, that *dispersion is the natural means to healthful concentration*'.

> As population spreads into the remote interior – the true mine of our national wealth – new townships spring up spontaneously; these again, as their inhabitants increase, especially as they increase by births, will contribute to still wider penetration into 'the regions beyond'; and thus, by an easy but vigorous system of action and re-action, diffusion and centralization will be reciprocally and simultaneously promoted, and by their combined operations, will transform this splendid wilderness into a series of domestic circles, and of large and flourishing communities. (Mansfield, 1841: 28)

There was remarkably little, in the actual circumstances of pastoral expansion, to support this hope. Economic historians have generally borne out Glenelg's more sober diagnosis, that inland Australia was destined to become a country of large flocks and small, widely scattered towns (Butlin, 1964b: 184–93; McCarty, 1978; Frost, 1991; Statham, 1989).[6] Only if something unexpectedly disrupted these expectations were the inland towns likely to enjoy better fortunes.

WATERING

The discovery of gold in New South Wales in 1851, swiftly followed by even richer finds in the new colony of Victoria, was the most decisive shift in the pattern of Australian settlement in our entire history. While grazing had reinforced the principle of dispersion, goldmining was a powerful, if volatile, force for concentration. Within a very few years, it had created dozens of entirely new towns, often in places remote from previous pastoral and agricultural settlement. These 'instant cities' and 'instant towns', places like Ballarat and Bendigo, were by far the most populous non-metropolitan centres in Australia, and remained so well into the 20th century. They were also wealthy beyond the expectations of most agricultural and pastoral communities. At the end of the 1850s, the non-metropolitan urban share of the Australian population had reached an all-time high. In Victoria, the centre of the gold rush and the most populous colony in 1861, the goldfields drew more people than either the capital city or the rest of the hinterland.

With the growth of the gold towns, Australia was introduced, in the most dramatic possible way, to the problem of sustainability. Gold, as everyone knew, was a precious but wasting resource. The miners who rushed to the fields were mostly young single men, likely, when the gold ran out, to rush off again to some other instant town. Their mobility and avarice seemed, to some contemporaries, to challenge the very notions of the gold town as a 'community', rather than an aggregation of egoistic individuals. Mining towns, moreover, were by their nature impermanent. Everyone who grew up in them knew that sooner or later the mineral would be worked out and, unless it found a new economic base, the town would wither and die. Their extractive and exploitative character was written on their topography in unsightly mullock heaps and denuded forests. By the 1870s, the long process of decline was already under way (Hall, 1968: 53). Over the following century, 'the rush that never ended' (Blainey, 1963) created many more instant towns in places ever more remote from the capital cities. In the 1880s, the mining frontier had reached outback New South Wales, northern Queensland and the Kimberley; by the 1890s, it was planting new towns in Western Australia. All these towns went through a similar cycle of boom and bust, rapid rise and drawn-out decline.

In spite of their insecure foundations, these towns often acquired a desire for permanence. Even when the gold began to give out, people stayed on, creating new manufacturing industries to support the local economy. The impressive public buildings and public parks of Ballarat and Bendigo, as well as smaller towns like Castlemaine, Bathurst, Gympie and Broken Hill, attest to the locals' zeal for 'community-building'. Their often skilled and well-educated populations built a large stock of 'social capital'; the gold towns were hives of voluntary activity. The political clout of the goldfields towns made them powerful claimants of state patronage and public employment; they became the favoured sites for the establishment of centrally-funded hospitals, lunatic asylums, gaols and technical institutes. Many of the political movements that emerged from the gold era, such as the demand for the free-selection of crown lands, were inspired by an aspiration towards the creation of sustainable rural communities. The distinctive Victorian tradition of Liberalism, based on tariff protection and free selection, had at its core the belief that the resources of the state should be used to secure independence and prosperity for its citizens. Many of the gold-rush immigrants came to Australia imbued with agrarian ideals, some-times born of Irish poverty or English radicalism. In 1854, William Westgarth observed the growing demand for agricultural settlement:

> ... because the natural longings of a great portion of the people are thus gratified, and because the face of the country has a pleasantly settled aspect, while its transactions rest on a broad basis of *perma-nent prosperity*. (Westgarth, 1854: 154; emphasis added)

Permanence and prosperity may have been more important than ideas of community to the land reformers of the 1850s and 1860s. By 'unlocking the lands' held by the squatters for free selection, they sought, first of all, economic and political 'independence' for them-selves and their families (see, for example, Macintyre, 1985: 19–39). Colonial Liberals held strongly to ideals of individual enterprise, but believed that, especially in the pioneering phase of development, the state might assist in creating the conditions under which individuals and communities could prosper. Political economist WE Hearn noted:

> It is a necessary incident of the imperfect stage of political develop-ment that pertains to a very young country, that the Government is obliged to undertake many functions from which at a more advanced period it is relieved. (Civil Service Commission, 1859–60)

This belief was especially strong in Victoria, where the challenge of find-ing permanent livelihoods for the gold-rush generation was most acute, and where Liberals supported state provision of railways, telegraphs, schools and irrigation projects. The need to provide elementary educa-tion, especially to rural communities too weak and dispersed to provide it for themselves, was one of the forces impelling the creation of centralised, state-funded education systems in the 1870s and 1880s.[7]

Historians have traditionally contrasted the weakness of Australian rural communities with the strength of American 'small-town democracy'. New Zealander William Pember Reeves observed in 1902:

> In rural Australia we saw scattered settlers living peacefully in thinly-spread or semi-nomadic provinces. Local life and organisations have never been intense. ... Unarmed, loosely-knit, not oppressed, the Australian country settlements were easily influenced by the central authority. (Reeves, 1969: 62)

According to this view, Australia's tradition of 'big government', like America's tradition of 'small government', was shaped by the nature of its rural frontier. Geography alone did not determine the Australian political tradition, but there is little doubt that rural communities were among the strongest claimants upon it.

Thirty years after the gold rush, when the dreams of many free selectors had been crushed by squatter resistance, drought and lack of capital, the ideal of rural independence remained strong. Since nature limited the potential for closer settlement, some nationalists argued that it was imperative to overcome the shortcomings of nature by irrigation. After a visit to California in 1885, the young champion of Victorian Liberalism, Alfred Deakin, predicted that:

> ... if Victoria is to continue to progress in the settlement of her people upon the lands and the multiplication of her resources by the conquest of those areas hitherto regarded as worthless; if she is to utilise her abundant natural advantages, bring her productiveness to the highest point, and secure to the agricultural population of her arid districts a *permanent prosperity*, it must be by means of irrigation. (Royal Commission on Water Supply, 1885: 779; emphasis added)

A century after the first European settlement, Australians continued to believe that human ingenuity might yet enable the drylands of Australia to flower as the rose. Providing irrigation would be a responsibility of the state, but this would merely create the conditions for individual and community enterprise. Irrigation promised to create communities based on more intensive agriculture, with the potential to support robust and vibrant community institutions. To some extent, pioneers of irrigation succeeded, although as early as the 1890s, when a long drought gripped the inland, there were intimations that the promised prosperity might be less sustainable than they had hoped.

PROTECTING

In the first quarter of the 20th century, Australians began, for the first time, to question the sanguine assumptions of 'Australia Unlimited' (Brady, 1918). The geographer Griffith Taylor was the best-known critic of the view, still held by many patriots, that the continent could support a population of 50 or 100 million (Powell, 1988, 1993). Environmental realism was reinforced by the new and sobering recognition that it was not only the mining towns that were in slow decline; so, too, were many of the recently settled farming regions of Australia. As early as the 1880s, South Australians had learned a chastening lesson on the limits of agricultural growth, when thousands of wheat farmers, lured beyond 'the margins of the good earth' in the previous decade, were forced by drought to beat a humiliating retreat (Meinig, 1970: 78–92). By the 1920s, scarcely a generation after the first settlers had cleared the Mallee and Wimmera, drought and mechanisation were accelerating the movement of population from the wheatlands of Victoria and South Australia.

Patriotic Australians were worried about a national malaise that they called 'the drift to the metropolis'. They observed the steady growth of the capital cities and the relative decline of the countryside. Between 1901 and 1954, the population of rural Australia actually remained more or less constant at around two million people, although as a proportion of the national population it fell from around 40 per cent to barely half that proportion (21 per cent). The population share of the non-metropolitan towns also declined, although more slowly and unevenly. From around 25 per cent of the population in 1901, it fell to around 18 per cent in 1947, but briefly revived during the postwar period to 25 per cent in 1954. Meanwhile, the capital cities had grown from about 1.2 million (or 31 per cent of the population) in 1901 to 4.8 million (or 54 per cent of the population) in 1954. Since the rural areas generally had higher fertility rates than the cities, much of the growth of the metropolitan areas was indeed founded on the migration of country people, especially young people, to the coastal cities (Merrett, 1977: 23–26, 1978: 190–91). One of the important effects of this migration was that many city folk continued to feel a close kinship with their country parents, siblings and cousins, a link that may have helped to sustain 'countrymindedness' as a national ideal (Aitkin, 1988: 50–58).

Rural decline was more pronounced in some regions than in others. Victoria, where the decline of the gold towns dramatised the 'drift of population to the metropolis', appointed a committee of inquiry to investigate the problem in 1916. The Government Statist, Arthur Laughton, identified old mining areas as the main sources of the exodus, offset to some extent by closer settlement in dairying and irrigation areas. The migrants, he believed, were drawn by the advantages of city life, better wages, more regular employment, more regular hours of labour and more social advantages, including opportunities for educating children and starting them in life (Laughton, 1916).[8] Other witnesses added to the list of causes: droughts, high tariffs, technological changes in agriculture and high transport costs.

Many feared that the drift of population to the cities presaged a decline of the nation itself. 'The aggregation of cities is like a wen, which draws an injurious sustenance at the expense of the general vitality', warned the Victorian rural paper, the *Leader* (1918). The Great War, it was believed, had demonstrated the military superiority of the bushman, and that the nation's defence, as well as its economic survival, depended on the continued vitality of the countryside. 'The best crop on our farms is the annual crop of babies', a 1920 New South Wales Select Committee on Agricultural Industry reported:

> When one realises that any policy which will multiply our farms must also increase the source of our best population, then surely it will be recognised that it is worthwhile to make it a national aim to obtain that policy and put it into operation. (Select Committee on Agricultural Industry, 1920: iii)

Soldier Settlement, the scheme launched by the Commonwealth government to enable ex-servicemen to settle on the land, was designed 'to help the soldiers become a stout yeomanry' and, in the process, to promote decentralisation (Lake, 1987: 24; Garton, 1996: 118–42).

Sustaining rural communities was therefore a national goal that transcended market forces. Within the national consensus that Paul Kelly has called 'The Australian Settlement', rural Australians were entitled to a form of social protection analogous to that afforded to urban workers through the Arbitration system (Kelly, 1992: 1–16). Just as the protective tariff and arbitrated wages were to guarantee workers a 'living wage', so farmers and their families should be afforded the means of obtaining a decent livelihood:

> Surely it is obvious that while country conditions are such that men
> will only go there when they are driven, instead of remaining there

by choice, no solution to the difficulty can be found. The average man brought up in the country leaves it because in the first place he gets tired of the drudgery, the monotony, the constant struggle against difficulties, and in the next because he can make a living more easily in the city. If it is less healthful at all events it does not involve a never-ending struggle against disadvantages that are sometimes almost heart-breaking. The first thing to be done is to see that every man who goes on to the land, if he has the necessary capital and applies the necessary skill and industry, shall be ensured a profitable return on his labour. (*Weekly Times*, 1918)

This social compact – the set of attitudes that political scientist Don Aitkin calls 'countrymindedness' – should include measures to make country living more attractive to the farmer's wife and children as well as more remunerative to the farmer himself (Aitkin, 1988: 51). Many contemporaries were impressed by the ideas embodied in American President Theodore Roosevelt's Commission on Country Life (1909; see Davison, 2003a). Modern transport and communications, medical care, rural education services and community centres would break down the monotony and isolation of country life, which bore most heavily upon country women. Distinct parties dedicated to advancing rural interests emerged first in Western Australia in 1913, and in all seven parliaments by 1920. By the 1930s, they had secured a number of concessions, including closer settlement schemes, subsidised railway freight rates, bounties and grants for agricultural products, as well as improved roads and telephone services in rural areas. The development of new state bureaucracies to foster rural development, and the stationing of their functionaries in country towns, reinforced the economic prosperity of the regional centres and added to their stock of social capital (Graham, 1966: 29–56).

Only occasionally during the interwar period did sceptics challenge the need to combat the drift to the cities. In 1929, New Zealand economist Allan Fisher contributed an article to the *Economic Record* arguing that most migrants from country to city were simply moving to better themselves. 'To complain about the world-wide drift to the towns is, indeed, to complain that the world as a whole is growing richer' (Fisher, 1929: 241). In 1942, another economist, WD Forsythe, challenged what he called *The Myth of the Open Spaces*:

Instead of crying 'back to the land' we have to get used to and make the best of an increasingly urban mode of life. The drift to the cities, misguidedly maligned as a source of social evils is in fact essential to rising standards of living. (Forsythe, 1942: 27)

These arguments made little headway in the 1930s and 1940s, but prefigure those of the 'economic rationalists' who began to dismantle rural protection in the 1970s and 1980s.

By the 1930s, drought and depression had given a new urgency to rural issues. Even in the coastal cities, great clouds of dust, swept from the parched and depleted farmlands of the inland, gave warning that all was not well in the Bush.[9] Perhaps the most perceptive contemporary observer of rural issues was Melbourne University Professor of Agriculture Samuel Wadham. An English-born soil scientist, Wadham also became a close and sympathetic student of rural communities (Davison, 2003b: 153–54). By the 1940s, his studies had led him to two strong, but potentially contradictory, conclusions. He was a strong supporter of social programs designed to improve life on the farm, and especially the lot of women and children, but his studies of soil erosion convinced him that much Australian agriculture was based on environmentally unsustainable foundations. 'Our farming systems have been ruining the land itself', he warned in 1943. Many of the programs designed to foster agricultural productivity, and in turn to assist closer settlement, were also accelerating the degradation of the land itself. So, while state assistance might be required to assist the farmer to stay on the land, state power might also have to be invoked to limit settlement and arrest the depletion of the precious and wasting resource on which farming was based (Wadham, 1943b: 15). It might be necessary to limit the farming population, as well as support it socially, in order to secure its future.

PLANNING

If drought and depression had compounded the problems of rural communities, the coming of the Second World War offered new hope that they could be solved. As the prospect of victory became more real, Australians began to focus their minds more intently on the New Social Order that would emerge after peace was declared. To fight the war, the nation had embraced more comprehensive measures of state control and planning. Many Australians were drawn to the comprehensive five- and ten-year plans followed by their Communist ally, the Soviet Union, and believed that both urban and rural communities should be planned in accordance with social ideals, rather than simply allowed to evolve through market forces.

The war itself had strengthened the appeal of decentralisation. The huge loss of life occasioned by the bombing of European cities, and the new threat of atomic warfare, convinced many postwar planners of the need to arrest the drift to the big cities. The reindustrialisation of the countryside was seen as the principal means of reversing the flow. Hydro-power in Tasmania, brown coal in Victoria's Latrobe Valley and the expansion of coal- and steel-based industry in the Hunter River, Illawarra and Lithgow, each anchored significant industrial developments in the immediate postwar period, the first since the turn of the century, when the non-metropolitan towns had increased their share of the national population.

Decentralisation was an ideal widely supported across the political spectrum (for a review, see Brown, 1995). Socialist publicist Brian Fitzpatrick regarded the concentration of industry in the capital cities as inimical to national defence and symptomatic of the concentration of capital in the hands of a few large private organisations. In order to counter it, he argued, the state should establish socialised abattoirs, wool-scouring plants, flour mills and freezing plants (Fitzpatrick, 1942). BA Santamaria, leader of the Catholic Rural Movement, sought to revive the rural community around the ideal of the self-sufficient family farm. 'There can be no Christian society which is not based on the solidarity and permanence of rural life', he told the Rural Reconstruction Commission in 1943. Farmers, he believed, were more independent of outlook than city-dwellers and more likely to resist the techniques of mass persuasion employed by totalitarian regimes. Healthy farming families required strong regional towns and the state should not flinch from supplying the resources to ensure their growth and vitality (Santamaria, 1943). Though their goals and assumptions differed, Fitzpatrick and Santamaria were at one in the conviction that rural communities benefited the nation and therefore deserved the support of the state. Like other advocates of decentralisation, they challenged the operation of the 'blind forces' of laissez-faire and were convinced that the social advantages of the city could be provided elsewhere 'if we plan for it and are prepared to pay the price' (Harris, 1948: 4).

In the 1950s, the renewed prosperity of the countryside, thanks to buoyant prices for Australia's wool and wheat, and the political successes of the Country Party in securing tariff protection for rural industries, may have disguised the fragile social foundations of rural communities. Mechanisation in the form of tractors, milking and shearing machines and combine harvesters was increasing the efficiency of agriculture and easing the labours of the farmer himself, but it was steadily lowering the demand for agricultural labour and, with it, the local employment prospects of the farmer's sons and daughters. Since the 1930s, the movement of young men and women from wheat-growing areas had accelerated (Holt, 1946: 40). In the postwar period, rural areas quickly took to the motor car and, as the barriers of distance collapsed, so too did small towns lose out to larger provincial centres. These trends were most marked in the dryland wool and wheat areas of western Queensland, New South Wales and Victoria and the northern districts of South Australia, although they were offset somewhat by the growth of population in irrigation areas and in some non-metropolitan industrial areas like the Hunter and Latrobe Valleys.

Gradually, however, the economic, social and political pillars supporting rural communities were beginning to weaken. During the long reign of the Liberal–Country Party coalition (1949–72), Country Party leader and Minister for Trade John McEwen became the architect of a wide range of protective measures, including tariffs and marketing schemes such as the Wheat and Wool Boards, designed to assist rural industries and communities. However, already by the late 1960s, some Liberals had begun to question these arrangements, and in 1972, when the Whitlam Labor government took office, one of its first acts was to introduce an across-the-board tariff cut. Rural and mining industries still accounted for a large share of Australia's exports and the regions where they were located elected more political representatives than their population alone would have entitled them to. But by the 1970s, city people were less inclined to concede the old moral and eugenic arguments in favour of rural life. They no longer regarded the crop of country babies as biologically or morally superior to their own. Fears of nuclear war or bombardment from the sea no longer made them worry about excessive concentration in the capital cities. The Whitlam government's plan to create new growth centres at Albury-Wodonga and Noarlunga was the last attempt to use federal fiscal and planning powers to offset the concentration of population in the coastal cities. Its failure sounded the death knell of decentralisation as an official goal of state policy in Australia (Lloyd & Troy, 1981). By 1985, Don Aitkin could conclude: 'In my judgement countrymindedness is finished as an ideology, even though its institutional and administrative arrangements will continue indefinitely' (Aitkin, 1988: 56). Even he can have little anticipated how soon those arrangements too would be swept away.

RURAL CRISIS OR JUST MORE OF THE SAME?

For 200 years, European Australians have struggled to create prosperous and sustainable rural communities. In the early colonial period, the sparsity of rural settlement was viewed as an infantile phase of development that would be overcome as the country matured. By the end of the 19th century, however, it was clear that 'permanent prosperity' in the countryside would require more vigorous action by the state. Fifty years later, after several decades of steady cityward migration, and the apparent failure of tariffs, rail freight subsidies, rural medical and education services and closer settlement schemes to stem the flow, many Australians had begun to wonder how many of their rural communities were sustainable at all.

During the government of Malcolm Fraser (1975–83), rural policy became one of the main battlegrounds between economic liberals (the so-called 'Dries') and those, like National Party (formerly Country Party) leader Doug Anthony, who fought for the retention of tariffs. By the late 1970s, however, the intellectual case for the protection of rural industries had been all but abandoned, even among sections of the rural population itself (Kelly, 1992). As its political base shrank, the influence of the National Party within the Coalition also gradually declined. From 19 of the 121 seats held by the Coalition in 1949, it shrank to 18 of the 148 in the first Howard government in 1996. As we have seen, the 'Dries' were not the first to question the use of state power to support rural industries and communities, but in the 1980s and 1990s, when neoliberal doctrines came to dominate public policy and Australia was opened up to global markets as never before, rural communities were among the first to feel the effects.

Taking a long view, the 'rural crisis' of the 1990s may be regarded as no more than the culmination of processes under way for a century or more. Almost none of the forces that produced it – technological change, environmental stress, collapsing international markets, rising debt – was entirely new, and most of its effects – rural depopulation, economic hardship, loss of community morale – had been experienced in previous times. What was new was the strength of the combined force with which they now acted, and the changed framework of expectations in which their impact was now interpreted. Some parts of the countryside continued to prosper and the non-metropolitan share of the national population even rose (Hugo, 2003). The migration of young people from declining rural and non-metropolitan industrial areas, such as the Latrobe Valley, was offset to some extent by the migration of retirees and others to the coast (cf. Department of Infrastructure, 1999: 7–13; Haberkorn et al., 1999). In the dryland farming areas, however, the decline of population and loss of young people, a process that has already been under way for several generations, accelerated further.

CONCLUSION

Any analysis of the current crisis that confines itself to short-term statistical measures of population movement, rural unemployment or income insufficiency misses an important but neglected factor. What generates the current sense of crisis is that country people are experiencing loss at a time when much of the rest of the country is prospering, when their community institutions are already debilitated by several decades of change, and when nobody seriously proposes that the situation can be reversed. With hindsight, we may judge that most of the previous phases of rural development in Australia were founded on false hopes. Rural communities would not grow of their own accord, as the early colonial administrators had expected. Even measures to protect and support them in the interwar period, or plan for them in the 1940s and 1970s, had limited success. But in each of those previous eras rural communities could hang on in the belief that the rest of the country needed and wanted them, and that theirs was a national cause. Now, for the first time in more than two centuries, that sustaining belief has been discarded, effectively if not explicitly. The ethic of development on which rural progress had long rested is now challenged by an ethic of conservation that portrays many traditional farming practices as harmful and imposes a new obligation of 'sustainability'. Meanwhile, the sense of community cohesion in rural areas is broken down by the impact of global markets and central government demands that local services should be contracted out, often to companies from out of town. The current crisis in rural Australia has been a long time coming, but free-market economics and environmental crisis are the hard rocks on which two centuries of rural hopes have finally foundered.

NOTES

1   Lockie (2000) briefly discusses the 'discourses of rural and regional Australia', but the historical context is thin. Gray & Lawrence (2001a) chart recent changes in rural ideology, and the 'de-traditionalisation' of farming, but provide only sketchy indications of how such ideologies and traditions evolved and where they came from. An exception in offering a longer-term perspective is Forth (2001).

2   For a collection of Wadham's work and a perceptive biographical portrait, see Blainey (1957).

3   Governor Macquarie was especially interested in town planning: see Atkinson (1996: 333); compare Macquarie (1956).

4   See the interesting discussion in Diamond (1997: 308–309).

5   Ritchie (1971) is quoting Macarthur to Bigge, 7 February 1821.

6   All these writers are concerned primarily to account for the dominance of the capital cities rather than the weakness of the inland towns, although as Statham suggests, once the capital cities gained a strong lead, through the centralising influence of convictism and pastoralism, the two tendencies were, to some extent, mutually reinforcing.

7   This is the view long held by educational historians. See, for example, Austin (1961: 112–14, 173–77). But note the objections to this view of John Hirst (1967: 42–59) in relation to South Australia.

8   Perhaps because of wartime economies, the evidence to this Committee was not published in the customary way in *Parliamentary Papers*. The Committee's brief report was published as *Victorian Legislative Assembly Select Committee on the Drift of Population* (1918: 641).

9   For the effects of the depression and other aspects of Victorian agriculture in the early 20th century, see Dingle (1984), chapter 9.

4

THE STATE OF RURAL POPULATIONS

Graeme Hugo

There are many myths that have been perpetuated about non-metropolitan Australian communities and their residents, but among the least tenable is stereotyping them as being less dynamic and less differentiated than the nation's metropolitan sector. This chapter seeks to dispel such myths by examining some of the contemporary and impending dynamics of the non-metropolitan sector through the window of shifts in its population. The most important resource in regional Australia is its people, but our knowledge of them is somewhat limited. It may come as a surprise that 35.2 per cent of Australians lived outside of cities with more than 100 000 inhabitants at the beginning of the 21st century, and they are changing in substantial and important ways under the influence of economic, social, political and environmental shifts. This chapter summarises trends in the size, composition and spatial distribution of this non-metropolitan population. It argues that this population is becoming more diverse and that this increased diversity offers considerable potential in efforts to work toward social, economic and environmental sustainability in rural Australia. At the same time, there has been a polarisation within non-metropolitan Australia that presents major challenges to policy-makers.

DEFINING NON-METROPOLITAN AREAS

With the burgeoning of interest in regional issues in Australia, there has been a great deal of confusion about who should be included in considerations of that sector. Part of the confusion has arisen from a lack of conceptual clarity. Terms such as 'regional', 'rural', and 'remote' are employed, sometimes with a specific meaning and in other cases more vaguely. Much of the present confusion stems from an attempt to combine into a single classification two distinctly different conceptual elements:

- urban/rural
- accessibility/remoteness

It is argued here that these are quite different concepts and need to be treated as such in differentiating types of settlement. An area can be both urban and remote or rural and remote. Any attempt to classify non-metropolitan Australia into rural and remote areas is misplaced. We need to classify areas in terms of their urbanness/ruralness and we also need to classify them by their degree of remoteness. The Australian Standard Geographical Classification (ASGC) has a system of identifying and classifying urban and rural areas and has recently incorporated a classification of accessibility/remoteness that allows areas outside of the major cities of the nation to be classified according to their degree of accessibility to services. This classification, known as ARIA (Accessibility/Remoteness Index of Australia), has been developed in the National Key Centre for Social Applications of Geographical Information Systems (GISCA) at the University of Adelaide (Bamford et al., 1999).

The definitions employed in this chapter can be clearly stated. The basic building block in the ASGC is the Census Collection District (small spatial units with 200–300 households). These are aggregated into a range of spatial units. From the perspective of the present study, the first distinction is between urban and rural. Urban areas are defined using a complex set of rules about population density and size (Hugo et al., 1997) and rural areas and populations are obtained as a residual. Part of the ASGC structure are 'Sections of State' as follows:

- Major Urban – urban areas with a population of 100 000 and over
- Other Urban – urban areas with a population of 1000 to 99 999
- Bounded Rural Locality – rural areas with a population of 200–999 population
- Rural Balance – the remainder
- Migratory – areas composed of offshore, shipping, and migratory CDs

The 'other urban' and 'rural locality' sections are at times grouped in this chapter as 'country towns'. The term 'non-metropolitan' is used to refer to all parts of the country outside of the major urban centres with more than 100 000 inhabitants. However, we can also differentiate non-metropolitan areas according to their degree of remoteness. ARIA indices of remoteness have been calculated for 11 338 localities outside of Australia's major cities and the entire area of non-metropolitan Australia has been classified into five categories of remoteness:

- Highly Accessible – locations with relatively unrestricted accessibility to a wide range of goods and services and opportunities for social interaction
- Accessible – locations with some restrictions in terms of accessibility to some goods, services and opportunities for social interaction
- Moderately Accessible – locations with significantly restricted accessibility to goods, services and opportunities for social interaction
- Remote – locations with very restricted accessibility to goods, services and opportunities for social interaction
- Very Remote – locationally disadvantaged, with very little accessibility to goods, services and opportunities for social interaction

Our concern here is with the sectors of Australia that are outside the 'major urban' and 'highly accessible' categories.

POPULATION TRENDS IN NON-METROPOLITAN AUSTRALIA

Table 4.1 indicates the changes that have occurred in the numbers of places and in the population in different size categories of places over the last three decades. It will be noticed that there was a substantial increase in the number of non-metropolitan (or country) towns, from 450 in 1966 to 728 in 1996, while their share of the national population increased from 21.4 to 23.7 per cent. However, between 1996 and 2001, there was a *decline* in the number of country towns from 728 to 692 (21.3 per cent of the population). Moreover, it will be noticed that the bulk of this decline was in the smallest urban category (1000–1999 inhabitants), indicating that many centres in this part of the urban hierarchy are experiencing overall population losses. The proportion of Australians living in rural areas (that is, localities with less than 1000 residents) also fell from 16.9 per cent in 1966 to 13.7 per cent in 2001.

Table 4.1 Distribution of population by settlement size, 1966, 1996 and 2001

	Number of urban centres			Percentage of population		
	1966	1996	2001	1966	1996	2001
500 000 and over	5	5	5	56.0	53.1	54.0
100 000–499 999	4	8	9	5.4	9.2	10.8
20 000–99 999	22	50	43	6.8	9.8	8.8
2000–19 999	250	366	364	12.4	11.3	10.4
1000–1999	178	312	285	2.2	2.5	2.1
Total urban	459	741	706	82.9	86.0	86.3
Total rural	n.a.	n.a.	n.a.	16.9	14.0	13.7
Total population (a)	n.a.	n.a.	n.a.	100	100	100
Total number ('000)	n.a.	n.a.	n.a.	11 599	17 892	18 972

(a) Includes migratory population.

Source: Rowland (1982); Australian Bureau of Statistics 1996 and 2001 Censuses

In 2001, 46 per cent of Australians lived outside the five largest cities and 34.5 per cent lived in non-metropolitan areas; that is, outside cities of 100 000 or more residents. The total non-metropolitan population amounted to 6.63 million people in 2001, compared with 6.68 million in 1996, 5.78 million in 1986 and 4.9 million in 1976. Hence, stereo-typing of regional populations as declining or static is incorrect. Nevertheless, population growth has certainly not been universal throughout non-metropolitan areas. The population in 'other urban' areas grew faster than 'major urban' populations during much of the three decades from 1966, although not in the 1996–2001 period. 'Rural' populations, however, have grown more slowly.

More importantly, there have been substantial regional variations among non-metropolitan areas in the patterns of population change. Figure 4.1 shows that between 1996 and 2001 there were distinct spatial patterns of growth and decline. The areas of population growth in regional Australia are strongly concentrated in certain areas, namely:

- the areas surrounding metropolitan areas
- along the well-watered east coast and south-west coast
- some resort and retirement areas
- some regional centres
- along the Hume Highway linking Sydney and Melbourne
- some relatively remote areas, especially those with growing mining activities, tourism, and significant Indigenous populations

Figure 4.1 Australia: population change, 1996–2001

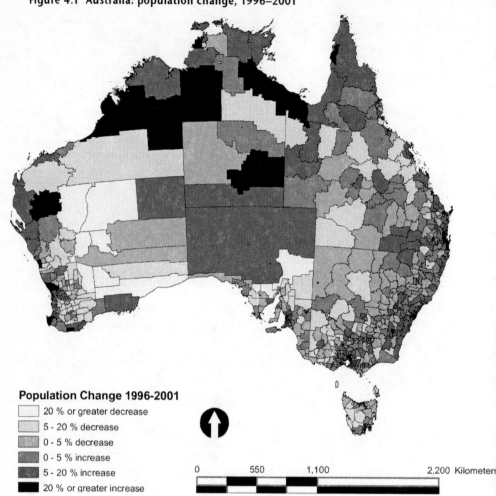

Population Change 1996-2001

- ☐ 20 % or greater decrease
- 5 - 20 % decrease
- 0 - 5 % decrease
- 0 - 5 % increase
- 5 - 20 % increase
- ■ 20 % or greater increase

0   550   1,100   2,200 Kilometers

Source: Australian Bureau of Statistics

On the other hand, there is also a spatial concentration of the areas experiencing population decline:

- above all, the dry farming areas of the wheat–sheep belt, such as in western Victoria extending through central-western New South Wales and Queensland, the south-east Eyre Peninsula and mid-north of South Australia, and the wheat–sheep belt of Western Australia
- many pastoral areas in central Australia
- certain mining areas, such as Broken Hill
- declining industrial cities, such as Whyalla in South Australia

These patterns point to a substantial degree of population variation within regional Australia.

It is interesting, too, to examine the patterns of population change in non-metropolitan Australia according to the degree of accessibility/remoteness of particular areas. Table 4.2 shows the rates of population change in the five accessibility sectors of non-metropolitan Australia. This indicates that only in the highly accessible areas close to major cities are population growth levels above the national average in both 1991–96 and 1996–2001.

Table 4.2 Australian non-metropolitan areas: population growth by level of accessibility, 1991–96 and 1996–2001

Level of accessibility	Rate of population growth (%)		Population density (persons per km^2)		Percentage of all farming families
	1991–96	1996–2001	1996	2001	2001
Highly accessible	6.2	6.6	77.2	80.0	5.9
Accessible	5.1	3.7	4.1	4.1	33.8
Moderately accessible	3.6	1.5	1.0	1.0	46.3
Remote	1.2	−1.0	0.2	0.2	10.7
Very remote	2.9	4.5	0.0	0.1	3.2
Total	5.8	6.0	2.3	2.5	100

Notes:

1   All calculations are made on the basis of SLA values aggregated to the accessibility categories.

2   The accessibility classification is based on the updated ARIA+ classification for the 2001 ASGC. Because the breakpoints between categories for the 2001 version have changed since the original ARIA, the breakpoints for the table above were adjusted to be concordant with the 1996–2001 table.

3   The population density data is based on Table B01 of the Basic Community Profile for 2001. The population change data is based on time series population estimates adjusted to the 2001 boundaries.

Source: Glover et al. (1999); Australian Bureau of Statistics 2001 Census; ABS (2003a: 46)

There is a decline in the rates of growth with increasing distance away from the large cities, except that the 'very remote' areas had a slightly faster growth rate than the 'remote' areas in 1991–96 and the 'remote' areas' population actually declined in 1996–2001. The 'very remote' areas were growing faster than other non-metropolitan areas in 1996–2001 because of their substantial Indigenous population as well as growth associated with tourism and mining activity. There is also an association between rates of population growth and population density and the areas with a high proportion of the nation's farming families also had low population rates.

Figure 4.2 Australia: population change in country towns, 1996–2001

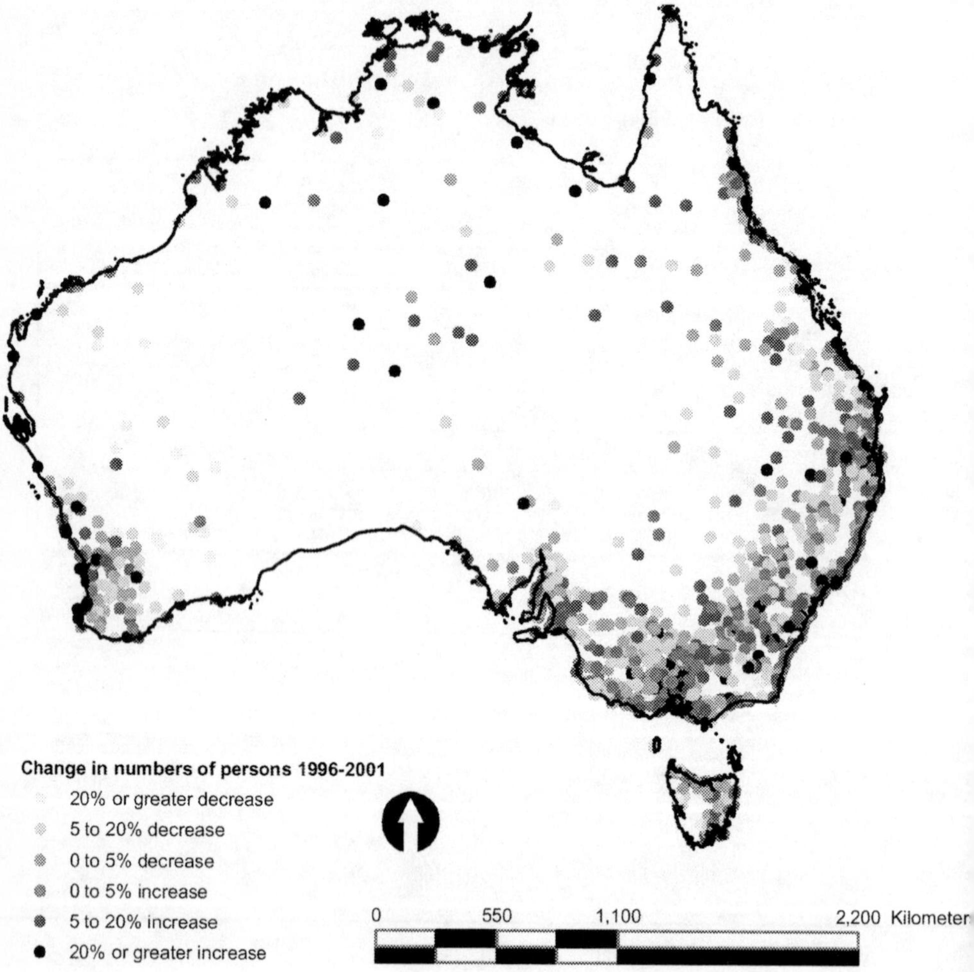

Change in numbers of persons 1996-2001
- 20% or greater decrease
- 5 to 20% decrease
- 0 to 5% decrease
- 0 to 5% increase
- 5 to 20% increase
- 20% or greater increase

0   550   1,100   2,200 Kilometer

Source: Australian Bureau of Statistics 1996 and 2001 Censuses

Turning to an examination of population growth trends in country towns, figure 4.2 shows the location of urban areas experiencing growth and decline. A clear spatial pattern is in evidence. Centres with relatively rapid growth are clustered around the nation's largest cities and strung along the eastern and south-western coasts. On the other hand, the wheat–sheep belt area tends to have urban places that are experiencing decline. In the more remote areas, there is a greater variation, with some centres experiencing growth and others recording decline.

In population growth and decline terms, there are strong indications of a growing dichotomy within non-metropolitan Australia. Holmes (1994) has argued that a dominant trend of the post-industrial era in non-metropolitan Australia has been a divergence between growing areas in coastal and metropolitan periphery areas and declining areas elsewhere. Analysing Census data up to 1996, Hugo and Bell (1998: 111) concluded that there is a growing dichotomy emerging in Australia's non-metropolitan areas, with eastern and south-eastern coastal zones and areas around the commuting sheds of each of the major cities growing at above national level, while in the heartland dryland farming and pastoral areas of rural and remote Australia absolute population decline is common and there is a consequent diminution in both their economic and social potential. The results from the 2001 Census indicate that this divergence has continued and intensified.

PROCESSES OF POPULATION CHANGE IN NON-METROPOLITAN AUSTRALIA

In examining patterns of population growth and decline, it is necessary to disaggregate the change and examine the processes influencing it separately. The demographic processes shaping population change in individual country towns, as in the non-metropolitan sector generally, are as follows:

- fertility – the extent to which women living in the area have children
- mortality – the pattern of death in the sector
- internal migration – the extent to which people move into the area from other parts of Australia and residents move elsewhere in Australia
- international migration – the extent to which people move into the area from overseas and residents leave for overseas

Taking fertility first, non-metropolitan Australia has shared in the national trend toward a lowering of fertility over the last few decades, but to a lesser extent. The non-metropolitan community is in general replacing itself through fertility, while the metropolitan population is not. On the other hand, mortality levels are higher in non-metropolitan than metropolitan areas. This applies to both infant mortality rates and expectation of life at birth.[1] There is a regular increase in mortality with increasing remoteness from Australia's major cities. This pattern is partly associated with an increasing proportion of the population being made up of Indigenous people with increasing remoteness: 38.3 per cent in very remote areas compared with 1.1 per cent in very accessible areas. Indigenous people suffer much higher mortality than the Australian population as a whole and this has some impact. However, it is also apparent that non-metropolitan people are at greater risk of death from some diseases and from accidents. In addition, their level of accessibility to health services is generally lower than residents in major urban areas and this contributes to higher mortality.

While there are differences in fertility and mortality between areas in Australia, it is differences in migration experience that are the main reason for variations in levels of population growth or decline. Migration differences are important, not only because they influence levels of population growth in areas but because migrants are never a cross-section of the population at either their origin or destination. The process of migration greatly influences the composition of the population of an area. The influence of migration can be differentiated between that of internal migration (within Australia) and that of overseas groups. By far the greatest influence is from internal migration (Bell & Hugo, 2000; Hugo, 2003) and this will be considered first.

An examination of the net migration[2] patterns for the non-metropolitan parts of each state between 1966 and 2001 reveals a number of interesting patterns. In the 1970s and 1980s, there was a substantial net loss of migrants from Brisbane, Sydney and Melbourne to their hinterlands. By the 1990s, however, this was reversed in Brisbane and greatly reduced in Sydney and Melbourne. The process of net loss from city to hinterland has been referred to as 'counterurbanisation' (Hugo & Smailes, 1985; Sant & Simons, 1993). It is apparent that counterurbanisation has decreased in significance in Australia since its peak in the 1980s. Hence, with respect to intrastate migration, there is a migration exchange between metropolitan and non-metropolitan areas favouring the latter only in the two largest states – New South Wales and Victoria – and this has been declining since 1986–91. There was a substantial reduction in the net losses from Sydney and Melbourne to their respective non-metropolitan areas in 1991–96 and 1996–2001. In Queensland, a net loss from Brisbane in 1986–91 was reversed to net gains in 1991–96 and in 1996–2001, while the net gains of Adelaide and Perth from their corresponding non-metropolitan areas increased in the 1990s. The Census data thus support the view that counterurbanisation trends have slowed considerably in Australia during the 1990s and have become more spatially concentrated.

It is important to point out that the movement between metropolitan and non-metropolitan areas is selective of particular groups. There is a substantial net flow from non-metropolitan areas into capital cities of young adults, related to the pattern of school-leaver age groups moving to large urban areas in order to obtain higher education or to gain a job in a much larger and more diverse labour market than is available in their home region. This pattern of out-migration of youth is a longstanding one in Australian regional areas (see, for example, Hugo, 1971). Similarly, the greater number of females than males in this movement has been a consistent feature over the last four decades. The fact that even fast-growing non-metropolitan areas cannot offer the post-school education and variety of job opportunities that are available in the capital cities means that these areas are also losing population aged in their late teens and early 20s. There is little chance that there will be any significant reversal of this trend toward a net loss of young adults. The key question is: To what extent can conditions be created to encourage some of these young people to come back to settle in non-metropolitan Australia after completing their education and gaining experience in metropolitan centres? This flow already exists. A long-term feature of migration in regional areas is a net migration gain in the late 20s and 30s age group. This is a function of the following elements:

- a return flow of some young people once they have completed their education in the city
- young people, especially women, some of them returnees moving in to marry a local
- in-movement of young adults (often families) as part of their career advancement in teaching, banks, stock and station agents, police, the health system etc. – the group known as the 'floating population'

These flows already exist and any attempt to increase in-movement to regional areas would attempt to encourage and enhance this movement.

One substantial change in the last decades has related to the 'floating population' referred to above. Studies in the 1970s (for example, Hugo, 1971) found that the movement of these groups into rural communities was crucial to their social capital. While the floating population stayed for only a short period before being transferred, they were often leaders in many local sporting, community, social, cultural and volunteer activities in country communities. Moreover, they were replaced by people with similar characteristics. However, with greater centralisation of services, the 'floating population' is increasingly moving between regional cities and not country towns. The result has been a significant loss to smaller rural communities, which is not only demographic but has a disproportionate impact on local social capital.

Another element in the inflow to some areas, particularly into coastal resort areas and some regional centres is the in-movement of retirees. As a result, the proportion of the population aged 65+ is greater outside of capital cities than within them. The more substantial ageing in non-metropolitan Australia is a function of two processes of importance:

- the out-movement of young people, which leaves the remaining older people as a greater proportion of the total population
- the in-migration of older people into resort-retirement communities and regional cities

As a consequence of these movements, these two types of areas had above average growth of their 65+ populations between the 1996 and 2001 Censuses.

We have dealt so far with the internal migration between metropolitan and non-metropolitan sectors, but it is apparent from an earlier analysis of regional population growth that there is considerable variation in the migration experience of different regional areas. Of the 50 non-metropolitan statistical divisions, only 16 recorded net internal migration gains during the 1996–2001 period. These were almost all located around major urban areas or along the east coast. Virtually all statistical divisions in rangeland and wheat–sheep belt areas recorded net internal out-migration.

One of the most dramatic changes in postwar Australian society has been the expansion of immigration and the diversification of the immigration intake. However, this transformation has been more marked in Australia's major cities than in regional areas. Over the 1947–2001 period, the number of Australia-born persons living in cities with 100 000 or more inhabitants more than doubled so that, in 2001, 59.9 per cent lived in such centres. On the other hand, the overseas-born population in the largest urban areas increased more than six times, to 82 per cent of Australia's overseas-born by 2001. Hence, the impact of immigration has been felt more in Australia's major cities than in the provincial cities or rural areas.

Over the 1947–2001 period, the proportion of the population in major urban areas made up by the overseas-born increased from 11.6 per cent to 29.2 per cent. Moreover, their impact upon the growth of those cities is underestimated by these figures, since the children born to overseas-born people after arrival in Australia are included with the Australia-born. The proportion of the total national overseas-born population living in provincial cities declined from 13.5 per cent to 10.8 per cent over the 1947–2001 period. In 2001, only 7.8 per cent of immigrants who had been in Australia less than five years lived in these centres. However, within provincial cities, the number of overseas-born increased almost fivefold and the proportion of residents who were overseas-born increased from 7.2 to 11.7 per cent. In rural areas, there was a substantial change. In 1947, a quarter of all overseas-born persons lived in rural areas, but this was drastically reduced to 7.3 per cent by 2001. In 2001, only 3.1 per cent of all immigrants who had been in Australia less than five years lived in rural areas. Nevertheless, the proportion of rural residents who were overseas-born increased from 7.6 per cent to 11.9 per cent. Hence, although the presence of overseas-born has increased in all three urban–rural sectors, the impact has been greatest in major urban areas.

Different overseas groups have differed in their tendency to settle outside of Australia's major urban areas, although all birthplace groups have a smaller proportion of their populations living outside large cities than the Australia-born. Most recently arrived groups, especially those from Asia, show a high propensity to settle in major cities and have only small populations in non-metropolitan areas. Exceptions include those from the Philippines, who are represented significantly in the 'other urban' and rural areas. This is partly due to the substantial number of Philippines-born women marrying Australia-born men in non-metropolitan areas. Also, some of recently arrived asylum seekers granted temporary protection have settled in rural areas, where they have worked in harvesting or abattoirs. Nevertheless, the overall picture is of overseas migration playing a much less significant role in the growth of non-metropolitan than of metropolitan populations in Australia.

THE INDIGENOUS POPULATION

While all overseas-born groups are under-represented in non-metropolitan Australia, this is not true of the Indigenous population. Special attention should be drawn to this group, not only because of their distinctive heritage but also because of their deprived economic and social circumstances. This is depicted in table 4.3, which compares the Indigenous and total population on a number of key indicators, drawing attention to important implications for provision of services.

Table 4.3 Australia: comparison of the Aboriginal and Torres Strait Islander population and total population, 2001

Characteristic	Aborigines and Torres Strait Islanders	Total population
Expectation of life at birth (years) – male*	56.3	77.0
Expectation of life at birth (years) – female*	62.8	82.4
Infant mortality rate*	12.7	5.3
Percentage in major urban	31.1	65.1
Percentage aged less than 15	39.3	20.8
Unemployment rate	20.0	7.4
Percentage employed as managers, administrators, professionals	15.6	28.0
Percentage labourers and related workers	24.7	8.8
Percentage with diploma, degree or higher	6.2	21.4
Individual income $20 748 or less per year	72.4	52.3
Individual income $52 000 or over per year	3.5	18.7
Percentage of households living in public rental accommodation	20.8	4.5

* 1999–2001

Source: ABS (2002b); Australian Bureau of Statistics 2001 Census

In 2001, 2.2 per cent of the national population was of Indigenous origin, but they comprised 4.3 per cent of those living in non-metropolitan areas and 1.0 per cent of those in major urban areas. A majority (68.9 per cent) of the nation's Indigenous population lived in non-metropolitan areas, compared with 34.9 per cent of the total population. Although the Indigenous population in major urban areas is increasing faster than that in the rest of Australia, they remain a predominantly non-metropolitan population and are an especially important group in country towns.

The Indigenous population is growing considerably faster than the total population, but around half of this growth in recent years has been due to an increased propensity of people to identify themselves as being of Indigenous origin (Gray, 1997). Nevertheless, levels of natural increase are higher among the Indigenous than among the non-Indigenous population. They differ substantially from the remainder of the population on all of the variables listed in table 4.3 and this reflects the highly disadvantaged position that they occupy in Australian society. Their needs remain a crucially important priority in consideration of non-metropolitan areas.

The non-metropolitan Indigenous population is distributed in a different way from the total population in the sector. They make up a major part of the population in the most remote non-metropolitan areas, in contrast to their much lower representation in the more closely settled non-metropolitan areas of the east coast, south-east and south-west. The highest representation of Indigenous people is in the Northern Territory, the northern three-quarters of Western Australia, northern South Australia, western New South Wales, and western and northern Queensland. Indeed, the spatial distribution of the Indigenous population is relatively even across the Australian continent and contrasts sharply with the highly concentrated total population distribution (Hugo, 2001).

CHANGING AGE STRUCTURE

One of the characteristics of populations that most shapes their pattern and level of demand and need for services is age structure. Hence, in looking to the future it is important to examine the changing age structure of the non-metropolitan population. Figure 4.3 overlays the age structures of Australia's metropolitan and non-metropolitan populations at the 2001 Census. It is strikingly apparent that the young infant and school age groups (0–14 years) are 'over-represented' in non-metropolitan areas, while young adults are 'under-represented'. This is a function of:

- out-migration of young adults from the non-metropolitan sector
- higher fertility in non-metropolitan areas than in metropolitan areas

Figure 4.3 Australia: age–sex composition of metropolitan and non-metropolitan population, 2001

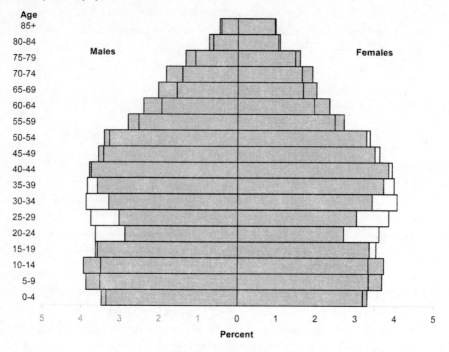

Source: Australian Bureau of Statistics 2001 Census

There is substantial 'under-representation' in the non-metropolitan age structure of young adults in the 20–34 age groups, reflecting the heavy loss of youth from non-metropolitan areas. As indicated earlier, there is also an 'over-representation' of the older population, due to retirement migration and the out-migration of young people. Yet ageing in Australia is frequently depicted, at least implicitly, as an urban issue. Clearly, regional ageing is of major significance. There is particular concern about the issue of the ageing of farmers. In fact, farmers work beyond the national retirement age of 65 more than any other group, with 15 per cent being aged over 65, compared to 9 per cent in 1986 (ABS, 2003a: 47). At the same time, the number of younger farmers has dwindled, with only 12 per cent of farmers being aged less than 35 in 2002, compared to 19 per cent in 1986.

However, it is important to point out that there is substantial variation within the non-metropolitan sector with respect to age structure. For example, growing non-metropolitan areas often have net gains in the young family formation ages and those with favourable ecological situations and substantial regional cities have net gains of the older, early retirement population ages. Areas of rural depopulation, however, often face significant problems due to ageing of their populations. The net migration losses are selective of young adults and consequently lift the average age of the local population. This often has social effects (for example, by reducing the number of sports teams in some country areas); this is often the key to a great deal of social activity in those communities.

THE CHANGING FAMILY IN NON-METROPOLITAN AREAS

Over recent decades, Australian families and households have become more diverse in their structure. The Australian Bureau of Statistics divides private households into family and non-family households, where the former category contains people related by blood or marriage and the latter includes group and single-person households. In metropolitan areas, 69.5 per cent of households are families, but this increases to 69.8 per cent in rural localities and 76.4 per cent in rural areas. On the other hand, 'other urban' areas now have 67.8 per cent of their households in families. This represents a substantial change from 1996, when 72.4 per cent were living in families and indicates the increasing numbers living in single-person and group household situations in provincial centres and larger country towns. In fact, the number of one-person households increased by 18.9 per cent between 1996 and 2001 in regional cities and 17 per cent in coastal areas, compared to 11.2 per cent in metropolitan areas and 15.5 per cent in all non-metropolitan areas (Haberkorn et al., 2003, 67). This reflects the older age structure of these areas. Accordingly, average household size in non-metropolitan areas is smaller (2.3 persons) than in Australia as a whole (2.5 persons) and metropolitan areas (2.6 persons). Aged persons living by themselves do not necessarily experience isolation and loneliness, but this group makes up an increasing proportion of regional households and their needs must be recognised.

If we focus only on family households, table 4.4 shows that rural areas tend to have larger proportions of couple families (with or without children) than urban areas and a smaller proportion of single-parent families. It will be noted, however, that country towns have a higher proportion of one-parent families than do major urban areas. This is partly a function of many single-parent families moving to country towns from large cities to take advantage of cheaper, often public sector, housing (Hugo & Bell,

1998). This has important implications for service provision in many country towns, especially when non-metropolitan areas are often stereotyped as being dominated by 'traditional' family types.

In keeping with differences in family structure, there are some distinct differences in housing between non-metropolitan and metropolitan areas. In major cities, 78.7 per cent of Australians live in separate houses, but the proportions are greater in country towns (87.7 per cent) and rural areas (95.1 per cent). Moreover, half of rural families own their own home (51 per cent), compared with 40.9 per cent in country towns and 40.5 per cent in major cities. The low proportion of

Table 4.4 Australia: family type by section of state, 2001

	Major urban	Other urban	Rural
	(%)	(%)	(%)
Couple families	82.1	81.3	88.4
Couple only	34.3	38.3	39.1
With dependants	38.7	36.5	41.0
Other (a)	9.1	6.6	8.3
One-parent families	15.8	17.3	10.6
With dependants	10.6	13.1	7.6
Other (a)	5.2	4.2	3.0
Other families (b)	2.1	1.4	1.0
Total	100	100	100
	'000	'000	'000
Total families	3212.80	1046.90	642.9

(a) Comprises families with non-dependent children and/or other relatives only.

(b) Comprises families of related adults, such as brothers, sisters, aunts, uncles.

Source: Australian Bureau of Statistics 2001 Census

Table 4.5 Australia: changing employment patterns, 1976–2001

Total employed persons	N U M B E R			
	1976	1981	1991	1996
Major urban	3 830 365	4 074 588	4 588 113	4 943 741
Other urban	1 121 946	1 298 271	1 450 393	1 612 620
Rural	1 028 036	913 055	1 062 925	1 072 328

Source: Australian Bureau of Statistics 1976, 1981, 1986, 1991, 1996 and 2001 Censuses

home owners in country towns and regional centres reflects the higher turnover of population than in the metropolitan centres or in more fundamentally rural areas. This is partly due to the presence in these centres of a large 'floating population' of employees on transfer by state and federal government departments, such as the post office and education department, private firms – particularly banks, stock and station agents and, increasingly, supermarket chains (Hugo, 1976: 67).

Housing costs represent a significant difference between urban and rural areas. Not only were there lower proportions of rural dwellers renting (15.9 per cent, compared with 29.6 per cent in towns and 28.9 per cent in cities), but, compared with residents of major cities, they paid less in weekly rent (median $100–149, compared with $150–199) and on monthly mortgage payments ($600–799, compared with $800–999). The median weekly rent and monthly mortgage payments of families in towns were similar to those of rural dwellers.

In accessing services, non-metropolitan households generally have to travel greater distances than their metropolitan counterparts. This is reflected in the fact that only 4.9 per cent of rural households are without a motor vehicle, compared to 11.5 per cent of other households.

CHANGING CHARACTERISTICS OF THE NON-METROPOLITAN POPULATION

One of the abiding myths of non-metropolitan populations is their supposed homogeneity. In fact, over recent decades there has been a consistent increase in diversity and a trend toward convergence with the metropolitan population with respect to many characteristics. An important area here is employment. Over the 1976–2001 period, there was an increase of 554 966 in the number of employed people living in non-metropolitan areas. Although this was only one-third of the increase in major urban areas (1 723 411), it is apparent from table 4.5

| | CHANGE | | (%) | | |
2001	1976–81	1981–91	1991–96	1996–2001
5 553 776	6.38	12.6	7.75	12.34
1 592 480	15.72	11.72	11.19	–1.25
182 412	–11.18	16.41	0.88	11.45

that the rates of increase in employment over that period have often been higher in non-metropolitan areas.

Employment change data from the Census are misleading in that they relate to the place of residence of the worker and not the location of the job, so that an important question remains as to the proportion of workers living in rural or other urban localities but commuting to jobs located within major urban areas. There have been increases in personal mobility, which have made it possible for people to move out of large cities and live in rural areas at the edge of the city and well beyond it, and yet still work and engage in other activities in the major urban area. This has contributed to a blurring of the characteristics of metropolitan and non-metropolitan populations in Australia. This convergence has also been facilitated by greatly increased mobility of the 'native' non-metropolitan population. This enhanced mobility has also been a significant factor in the retention of non-metropolitan populations and has extended the area over which farm dwellers can range in seeking off-farm employment.

From an examination of the changing mix of industry types providing employment in the metropolitan and non-metropolitan areas, we can establish what kinds of jobs are being created and lost in non-metropolitan areas. Table 4.6 shows changes between the 1986 and 2001 Censuses with respect to employment. The structural changes occurring in the economy as a whole are immediately evident in that, although overall there was an increase of 25.3 per cent between 1986 and 2001 in the total number of jobs, there were declines of 9.2 per cent in agriculture and 7.13 per cent in utilities. These were offset by gains of 46.3 per cent in trade, finance, administration, retailing and services, and 30.9 per cent in construction. The most striking differences are in the rural categories, where a decline occurred in the number of jobs in agriculture, the traditional mainstay of the rural economy. On the other hand, rural and other urban areas had a very rapid increase in trade, finance, property, business services, public administration, defence, community services and recreation. Clearly, there is a convergence occurring in metropolitan and non-metropolitan areas in their employment structures and an increased diversity in the workforce living in regional areas.

Accessibility to education services is a major issue in non-metropolitan Australia. Census data for 2001 show that the proportions of Australians in non-metropolitan areas with post-school qualifications are lowest in rural areas and highest in metropolitan Australia. Of course,

Table 4.6 Australia: percentage changes in employment patterns in urban/rural categories, 1986–2001

	Total	Major urban	Other urban	Rural
Agriculture	–9.22	41.6	29.08	–19.3
Mining and manufacturing	1.63	–1.29	0.15	26.3
Construction	30.91	35.68	16.88	32.64
Electricity, gas, water, transport, storage, communication	–7.13	–1.33	–25.22	–2.46
Trade, finance, property, business services, public administration, defence, community services, recreation	46.33	48.02	36.04	55.29
Total	25.25	33.29	21.28	22.44

Source: Australian Bureau of Statistics 1986 and 2001 Censuses

this partly reflects the out-movement of young people from rural areas to study in metropolitan areas and the fact that some rural-based jobs do not require post-school training. A greater proportion of families in rural areas have a primary reference person with vocational education than is the case in major cities. The largest difference is in respect to degrees and diplomas, which have a higher incidence in major cities.

There are, however, significant educational disparities in non-metropolitan Australia and a major issue relates to differential access to educational opportunities compared with those enjoyed by children in metropolitan areas. This is of particular relevance when it comes to secondary education. Whereas 87.4 per cent of metropolitan-based 16-year-olds attend school, this is the case for only 82.7 per cent of those living in non-metropolitan areas; in remote areas, the proportion is 62.1 per cent. The proportion of the population who completed year 12 decreases regularly with increasing remoteness (ABS, 2003b: 93). A similar pattern is evident in the numbers attending educational institutions.

INCOME

It is difficult to measure income in Australia and especially difficult to compare income between metropolitan and non-metropolitan areas because of the difficulty of accounting for some items, such as food grown and consumed on farms. Moreover, differences in the cost of living, as has already been demonstrated in relation to housing costs, make it difficult to compare the two types of areas. However, 1999–2000 Australian Taxation Office data indicate that the mean annual taxable income of people aged 15 years and over in non-metropolitan areas was $33 533 compared to $39 451 (17.6 per cent lower) in metropolitan centres (Haberkorn et al., 2003, 78). The proportion of households receiving a weekly income of $300 or less and qualifying as low-income households at the 2001 Census was 16.3 per cent in non-metropolitan areas and 12.8 per cent in metropolitan areas (Haberkorn et al., 2003, 79). However, there are considerable variations within the non-metropolitan sector. There is a concentration of low incomes in:

- many of the dryland broadacre farming areas, which in recent years have experienced low prices and droughts
- coastal areas, which have experienced a significant in-migration of retirees (who are hence low-income, although sometimes asset-rich)
- some remote areas dominated by Indigenous groups

There is even more of a contrast when high-income households (weekly income of $1200 or more) are considered. Some 34.2 per cent of metropolitan households fall into this category, compared to 21.4 per cent of their non-metropolitan counterparts (Haberkorn et al., 2003, 80). Higher-income households are concentrated around major cities and in some remote areas, especially those with mining communities, and some regional centres.

An increasingly important factor is migration. Hugo and Bell (1998) have shown that there is a strong economic selectivity factor operating in internal migration between metropolitan and non-metropolitan areas in Australia. The in-migration to large urban areas is selective of high-income groups; indeed there are net gains of high-income groups in these cities but net losses of low-income groups. On the other hand, there are net gains of low-income groups in non-metropolitan areas. This is a function of a substantial flow of low-income groups from metropolitan areas in Australia, partly associated with lower living costs (especially relating to housing) in many non-metropolitan areas. Part of the movement is retirement migration, but it also includes significant numbers of other income-transfer recipients. Accordingly, 36.5 per cent of the non-metropolitan population aged 16 years and over claim social security, compared with 29.5 per cent of the metropolitan population (Bray & Mudd, 1998: 8).

One of the major elements in the income picture in regional Australia is the increasing diversity of sources of income of farm families. In particular, farm families have become more dependent on off-farm income, especially in times of on-farm financial hardship. In 2000–01, off-farm income in broadacre and dairy farms was almost as much as on-farm income (ABS, 2003a: 47). Over the past two decades, there has been a substantial increase in both the level of off-farm income and its proportion of total farm income, especially in broadacre farming (Barr, 2000).

In 2001, the average weekly equivalised gross household income (that is, household income adjusted on the basis of the household size and composition) in major cities was substantially above the average in regional areas. Moreover, the increase in income over the 1996–2001 period was significantly faster in major cities than in regional areas. Turning to income distribution, the greatest inequality in Australia is found in remote areas, where clearly the low average Indigenous incomes contrast greatly with the high incomes of groups involved in mining and some tourism. On the other hand, there is less income inequality in other regional areas than in major cities. If we examine low-income households (those in the bottom 20 per cent of income earners), only 17 per cent of the residents of major cities were in low-income households, whereas the proportions were substantially higher in outer regional (27 per cent), inner regional (24 per cent) and remote areas (25 per cent). There was little change in this pattern between 1996 and 2001 (ABS, 2003a: 156). However, if direct housing costs are removed from income, the proportions of people living in low-income households are lower in outer regional and remote areas than in major cities. This not only reflects the relatively cheaper housing in these peripheral areas but also the very low quality of much of that housing.

There is a higher reliance upon government transfers in non-metropolitan areas than in major cities. In 2001, 39.4 per cent of Australian families received some government benefits, such as Newstart Allowance, Parenting Payment, Rent Assistance, Austudy and Disability Pensions (Haberkorn et al., 2003, 81). The proportion of non-metropolitan families (42.2 per cent) was higher than in metropolitan areas (37.6 per cent). Unemployment in Australia is also strongly associated with poverty and is lower in major cities (7.1 per cent in 2001) than outside them (8 per cent). However, the level of unemployment was highest in the more closely settled regional areas (8.5 per cent) and fell to 7.9 per cent in outer regional areas and 4.5 per cent in remote areas (ABS, 2003a: 126).

While discussions about poverty in Australia usually focus almost exclusively on the poor living in capital cities, it is important to realise that the incidence of poverty is higher among people living outside the capital cities. Socioeconomic disadvantage is not distributed uniformly across either urban/rural areas or areas differentiated according to their degree of remoteness. Whereas 62.7 per cent of Australians live in major urban areas, only 55.7 per cent of residents live in the most disadvantaged areas. Country towns (other urban and rural localities), on the other hand, have only 25.8 per cent of the population but 39 per cent of all areas in poverty. An interesting pattern is also evident in relation to degree of remoteness. Whereas the highly accessible parts of Australia have 81.8 per cent of the national population, they have only 74.7 per cent of those in poverty. There is, then, a clear concentration of the poor in non-metropolitan areas and within particular communities. The rural poor represent a fast-growing group in Australia deserving of policy attention.

CONCLUSION

Australia's non-metropolitan population is growing at least as fast as that in major cities and this will continue at least for the next decade. However, this growth is far from universal and, in fact, there is an increasing dichotomy between rapidly growing urban areas and those which are static or declining in population. Indeed, a significant amount of growth is occurring in towns at and just beyond the commuting limits surrounding major cities, so that some would argue that it is not so much a counterurbanisation trend that is occurring but more a new, diffuse form of urbanisation. Certainly, the sharp boundaries drawn by the Australian Bureau of Statistics around Australia's major cities are becoming less relevant with each passing year. Nevertheless, it is clear that all non-metropolitan growth is not of this type. This chapter has demonstrated that the non-metropolitan population is becoming more diversified. It is argued also that there is a widening polarisation occurring in non-metropolitan Australia. The rangelands are generally experiencing depopulation, dominated by school leavers; however, there are substantial areas in the better-watered and more accessible parts of non-metropolitan Australia that are continuing to experience significant and sustained net in-migration and population growth. The problems faced in the two different types of areas are quite different. Population change in non-metropolitan Australia is becoming, and is likely to become even more, diverse and perhaps less predictable than in the past.

With respect to the composition of Australia's non-metropolitan population, it is apparent that it is becoming more heterogeneous. There is an increasing tendency toward convergence of patterns and trends in the non-metropolitan sector toward those in large city populations with respect to many social, demographic and economic variables. However, the theme of divergence within non-metropolitan areas with respect to population growth trends presented above also applies to the crucial area of the wellbeing of the population. In the burgeoning of interest in non-metropolitan issues in Australia in recent years, much attention has been focused on a perceived gap in the levels of living of people living outside the major cities compared with the city dwellers. Such a focus, however, draws attention away from the crucially important and ever-widening gaps in wellbeing occurring between areas and groups *within* the non-metropolitan sector, just as is occurring within Australian cities. It is not only unhelpful to stereotype and generalise about the characteristics of people who live in non-metropolitan areas and the conditions that they experience, it can also substantially reduce the effectiveness of policies and programs initiated to overcome problems in non-metropolitan Australia. Interventions directed at all non-metropolitan areas will benefit those who are in need but also those who are not, so that their impact is diluted. There are groups and areas in country towns with urgent and important needs that must be addressed, but this does not apply to all country towns. Accordingly, interventions need to be carefully targeted so that their impact on the people and areas most in need can be maximised. The analysis presented here indicates that many communities in non-metropolitan Australia have populations that are younger and growing rapidly, with increasingly diverse and skilled labour forces and populations. This represents a substantial potential for the future in a context where geographical constraints on the location of economic and social activity are being loosened by developments in information technology and transport.

NOTES

1   The infant mortality rate (IMR) is defined as the number of deaths of children under one year per 1000 live births. Expectation of life at birth is the average number of years a person can expect to live if the current age/sex specific pattern of mortality is maintained throughout their lifetime.

2   Net migration can be defined as the difference between the number of persons who have changed their place of usual residence by moving in and the number who have changed their place of usual residence by moving out.

5

THE CHANGING SOCIAL FRAMEWORK

Peter Smailes, Trevor Griffin and Neil Argent

Chapter 4 demonstrates the diversity of social trends within non-metropolitan Australia, and suggests an increasing polarisation between demographic growth in the better-watered and more accessible rural areas, and a depopulating and declining remainder. Generalisations that can be made at the scale of an entire continent, however, may be less applicable in specific locales. In this chapter, working at the level of individual communities, we treat the reported spatial polarisation and increase in diversity as hypotheses to be tested within south-eastern Australia's rural heartland. Within this area, we investigate the relationship between individual rural communities as local social systems, and the environments – both physical and social – in which they are embedded. The prospects for sustainability of a local community depend not merely on the social capital and inner workings of the community, but also on the capital endowment and range of opportunities presented by the surrounding environment. We therefore differentiate between change taking place within the communities themselves, and change in their broader social and physical setting, and seek to discover how far these changes are leading to spatial polarisation. Although recognising the vital importance of qualitative data in understanding community adjustment to social, economic and technological change, the geographic scale of this study necessarily confined us to a quantitative analysis.

After defining our basic concepts and situating the study in time and space, we identify a series of key elements of the physical environment, and of a community's relative location in the space economy, that are likely to have a major impact on the economic and social functioning of a local social system, and that may change only slowly over time. Using the factors identified, we employ a clustering algorithm to reduce the 414 places studied, each with its unique set of environmental and locational qualities, to a manageable number (5) of environmental/locational types. Key characteristics of the clusters, and their spatial distribution are identified. We then identify some key characteristics of settlement distribution within each local community that are hypothesised to have a substantial impact on other important socioeconomic characteristics of the local population. These settlement characteristics can change more rapidly over time, and their actual changes over a 20-year period are demonstrated.

The hypotheses that diverging and polarising trends are emerging both between and within the clusters are then tested with respect to key socioeconomic and demographic variables. Finally, we draw out some implications of our findings for the detailed studies discussed in later chapters of the book.

THE STUDY AREA

The analysis was based on a crescent-shaped area that stretches from South Australia's Eyre Peninsula in the south-west, to the New South Wales/Queensland border in the north-east and as far inland as Lightning Ridge (see figure 5.1). This incorporates a large proportion of the Australian ecumene as defined by Holmes (1987). While excluding Tasmania, as well as important parts of the nation's 'sun-belt' in Queensland and Western Australia (O'Connor & Stimson 1996; Stimson 2001; Ward 1996), it nevertheless represents the entire gamut of rural locality types, from the remote and marginal farming/pastoral country of far western South Australia to the typical burgeoning 'sea change' communities of the New South Wales north coast (Burnley & Murphy, 2003). Major urban areas were excluded, along with most of the metropolitan fringe and penumbral zone of the 'city's countryside' (Bryant, 1992; Bryant et al., 1982). The New South Wales metropolitan areas in particular were broadly defined to exclude the whole central New South Wales coast from Port Stephens to Shellharbour (see figure 5.1). Also eliminated were contiguous uninhabited areas of 150 square kilometres or more, such as National and State Parks, state forests and major water bodies.

Figure 5.1 Defined spatial units and location of the study area

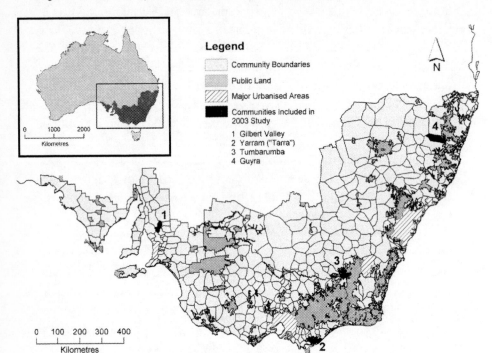

The 414 spatial units represent approximate social catchment areas, each centred on a significant country town. They were defined empirically for South Australia through detailed field and survey work, and for Victoria and New South Wales were approximated through the use of a model calibrated on the South Australian data. The method is described in detail elsewhere (Smailes et al., 2002b). Using look-up tables and constant boundaries for each Census, data at the level of Census Collection Districts (CCDs) were amalgamated, distinguishing in each case between the central town, any other identifiable clustered settlements, and the rural balance. The units in some cases correspond to local government areas, but seek as far as possible to represent place-based communities, each centred on a socially significant town, rather than official administrative or statistical districts. For convenience, these units are referred to as 'communities' in the sense of a local social system, though in the two eastern states these can be no more than approximations. Many of the catchment areas defined around the larger country towns will certainly have lower-level sub-systems – 'communities within communities'; detailed empirical research in South Australia has demonstrated the multi-level nature of place identification and social interaction (Smailes et al., 2002a). Nevertheless, this database allows a detailed study of the processes of social change in spatial units at a meaningful and manageable scale of resolution.

Data were collected from the 1981, 1996 and 2001 Censuses, but most attention here is on change over the full 20-year period. These two decades cover a period of rapid social and economic change, covering the opening of the Australian economy to the forces of globalisation by the Hawke and Keating governments, the international 'farm crisis' (Goodman & Redclift, 1989) and the restructuring processes that followed, and the adoption of neoliberal policies by the succeeding Howard/Costello governments, whose impact is dealt with in detail in other chapters of this book.

POPULATION, PRODUCTION AND ENVIRONMENT

In 1981, the defined study area already contained a population of 2.6 million people; perhaps counter-intuitively, by 2001 it had increased by over 20 per cent, to 3.1 million. The breakdown of this aggregate increase reveals a major change in the settlement pattern within rural communities. The 414 central towns included the non-metropolitan regional capitals, such as Bendigo, Ballarat and Orange. At the aggregate level, it is not surprising that they included over 60 per cent of the total population (almost 2 million in 2001) and showed a healthy increase over the period 1981–2001 (18.8 per cent). The most striking proportional change, though, was the doubling in population (from 143 000 in 1981 to 294 000 in 2001) of small subsidiary settlements, whose spatial distribution is discussed later. The balance of the rural population grew by about 13 per cent, to just over 850 000 in 2001.

The strong influence of Australia's natural environment on the location and concentration of settlement has been evident from the time of first settlement and formed the basis of Griffith Taylor's 'stop-go determinism' (Taylor, 1951), greatly affecting the distribution of primary industry (Wadham et al., 1964) and the location and dominance of metropolitan centres (Rose, 1966; Vance, 1970). Over the last half-century, and particularly in the timeframe of this study, the network of rural communities established on the basis of primary industries in the 19th and early 20th centuries (Powell, 1974, 1988) has been impacted by the twin forces of the international farm crisis weakening the farm-based component (Lowe et al., 1993) and the rising non-farm demand for rural space, resulting from counterurbanisation and the revolution in communications. The emergence in Australia of the 'post-productive countryside' (Argent, 2002) or, less radically, a 'multifunctional agricultural regime' (Wilson, 2001) has been moulded by a similar, though not identical, set of environmental factors to those originally significant for agriculture.

In a traditional productivist landscape of Australian rural communities, dependent on primary industries for their economic base, broad limits are set to the range and mix of viable farm enterprises by terrain (slope, altitude, local relief), climate (particularly rainfall amount, distribution and reliability) and soil quality, along with the availability or otherwise of irrigation water. These factors thus influence the size and density of the farm population, and some aspects of its composition. The same factors are also important in a post-productivist countryside, through their influence on perceived landscape and amenity values, though the strength of the relationships will vary. Locally, they may occasionally change sign: in growth areas, for example, hilly terrain of low productive value may produce scenery and recreational opportunity of high amenity value to an adventitious population (Smailes, 2002). Soil qualities are vitally important. Compared to large areas of North America and Europe, Australia's ancient geology and consequently nutrient-poor, thin and fragile soils have given rise to constant concerns over soil erosion and fertility depletion, and more recent alarm over the rapid spread of dryland and irrigated land salinity (Bailey & James, 2000; Sustainable Land and Water Resources Management Committee Working Group on Dryland Salinity, 2000).

The 414 communities were therefore each given scores expressing their rainfall, terrain and soil qualities, and access to irrigation water. While these variables (particularly soil qualities) are not constant over time, they are unlikely to change much over two decades.

A score on ability to make a living from land-based enterprises was also needed for each community. The productivity of the land is clearly more than the sum of its natural endowment, for it depends also on human inputs, market conditions, transport costs and much more. It varies from year to year depending on seasonal conditions, market prices, and input costs. To even out seasonal variations between towns and regions, the surrogate measure used for this variable was the average gross value of farm production per hectare for the farming years 1980–81 and 1995–96, expressed in constant 1989/90 dollars.

Once the 'bones' of the space economy (points of entry, inland resource concentrations, linking routeways, corridor intersections) have been laid out, relative location within a nation's space economy is of huge importance to a rural community's prospects for sustainability. While it does not fit readily into the five-fold classification of natural, produced, human, institutional and social capitals (see chapter 1, and Cocklin & Alston, 2003), it is nonetheless a vital arbiter of a community's economic and social potential. Each studied community was given a score on three spatial measures of relative location: the national 'ARIA-Plus' index of accessibility/remoteness from urban centres of varying sizes (Department of Health and Aged Care, 1999); a dummy variable expressing coastal/inland location; and the density of population at regional level surrounding the community, expressing accessibility to potential employment within an 80-kilometre commuting range. In addition, a temporal measure was used, since relative location also has a time dimension. A significant corollary of the colonisation process, spreading inland from a few coastal points of entry, is the overall duration of white settlement as a measure of the maturity and historicity of the (non-Indigenous) cultural landscape. The time of first settlement is strongly associated with the nature of the cadastre (size and shape of land parcels, layout and density of the road pattern, number of townships surveyed etc.), and with the presence of historic homesteads and buildings and other landscape features. This variable thus has substantial potential importance for farm size and type, and for perceived amenity. It was measured from the time of establishment of the main township in each community.

THE CLUSTER ANALYSIS

In terms of the 'capitals' framework for understanding community sustainability (Cocklin & Alston, 2003), the variables identified above represent a stock of 'capitals' endowed on any specific place-based community not by its own efforts, but by virtue of its position within the nation's physical environment (natural capital) and space economy. Each community's endowment is, of course, unique, but many show close similarities. To test our hypotheses of polarisation, heterogeneity and differential socioeconomic change in different types of environment, we first needed to reduce reality's infinite variety to a manageable number of types. A cluster analysis was therefore conducted, based upon each community's degree of similarity in relation to eight ratio-scale independent variables (figure 5.2).[1] The mean scores for each cluster on the eight input variables, and the incidence of the two binary variables, appear in table 5.1.

Figure 5.2 Classification of communities into five environmental/locational types

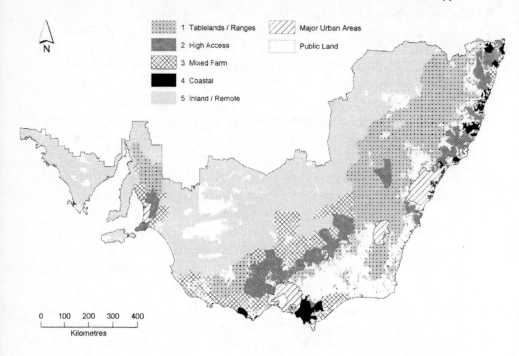

Table 5.1 Key environmental and relative location aspects of the five clusters

Environmental and locational variables	Cluster 1 Tablelands/ ranges (n. 110)	Cluster 2 High access (n. 72)	Cluster 3 Mixed farm (n. 74)	Cluster 4 Coastal (n. 36)	Cluster 5 Inland/ remote (n. 122)
Median annual rainfall (mm)	709	796	656	1236	402
Median altitude (m.a.s.l.)	482	299	95	125	130
Median slope index	1.80	1.37	0.41	1.51	0.28
Soil quality index	27.73	23.29	23.39	23.61	29.37
Mean value agricultural production ($ per hectare)	178	486	514	712	283
ARIA-Plus' remoteness index	2.83	1.01	1.73	1.49	4.98
Regional density of population ('000 in 75 km radius)	45	112	93	109	25
Duration of European settlement (years)	140	148	136	104	117
Number of coastal communities in cluster	8	9	25	21	24
Number of irrigation communities in cluster	0	2	21	0	35

Source: Data obtained by authors from a wide variety of sources

Cluster 5 (inland/remote) is spatially the most extensive, and occupies the bulk of the traditional dry farming wheat–sheep belt and the marginal lands between the ecumene and the rangelands proper. It is characterised by low annual rainfall, very low regional population density, flat relief and low elevation, and a relatively short duration of white settlement. While its average soil quality index is high, its average farm production value per hectare is low, and would be much lower except that it includes 35 of the 57 irrigation-based communities with intensive agriculture. It also includes about a quarter of the coastal communities, but most of these coastlands are relatively low and unspectacular, presenting sand dunes and low cliffs to the cold-water Southern Ocean and Spencer Gulf. Good harbours are few and river estuaries practically absent.

Cluster 1 (tableland/ranges) is the second most extensive cluster, occupying essentially the New South Wales tablelands and adjoining western slopes, along with the fringes of the Victorian high country, parts of the Wimmera and western Victoria, and the part of the South Australian mid-north that lies within Goyder's line of rainfall (figure 5.2). While including a good part of the higher rainfall dry farming belt, and also a number of pockets of orcharding and other more intensive land uses, this cluster has a very important pastoral component, reflected in a low average value of production per hectare despite the moderate to high average rainfall and soil quality index. The average altitude is the highest of any cluster, as is the average slope index, and the remoteness index is also high. Regional population density on average is almost twice that of Cluster 5 (inland/remote), but still very low compared with the other clusters. The few coastal communities are on the remotest parts of the New South Wales and Victorian coasts. There is no need for irrigation. This cluster includes three of the six case study communities discussed in the project team's earlier volume (Cocklin & Alston, 2003).

Cluster 3 (mixed farm) is much less remote than Cluster 1, closer to sea level and has lower average slopes. It has twice the regional population density and more than twice the value of farm production per hectare. This is influenced by the presence of a large number of irrigation-assisted communities, but most of these are supplementary to rainfed agriculture (as in north-central Victoria) rather than oases in an otherwise very dry environment (figure 5.2). Spatially, this cluster is rather fragmented, with four main contiguous groupings (north-central Victoria and the adjacent New South Wales Riverina, eastern Gippsland, the western Victorian coastlands stretching to south-eastern South Australia, and outer Adelaide Hills/northern Adelaide Plains). All these groupings are at arm's length from the metropolitan fringes, as is the scatter of nine places along the north coast of New South Wales that fall into this cluster.

Cluster 2 (high access) is generally closer to the metropolitan areas than Cluster 3, and has the lowest average remoteness, longest duration of white settlement, and highest regional population density of all the clusters. It has high rainfall, average soil quality and moderate altitude and slope, indicating a varied countryside. It produces a high-value agricultural output per hectare. Spatially, the main grouping stretches from central Victoria into southern New South Wales, following the Hume (Melbourne to Sydney) Highway along the flanks of the alpine ranges. A second important concentration is in northern New South Wales, east of the Dividing Range but in few cases reaching the coast itself; smaller groupings occupy the inner Adelaide Hills and the southern Shoalhaven country.

The final division is Cluster 4 (coastal), spatially the smallest cluster and occupying dominantly coastal or near-coastal locations in western Gippsland and the New South Wales north coast, with outliers on the Victorian Otway Ranges coastline, but unrepresented in South Australia. Reflecting their coastal location on the fringe of the ranges, the communities of this cluster have the highest average rainfall and value of farm production per hectare, and though less accessible on average than cluster 2, they still have a very high regional population density.

The five clusters represent a reasonable division of the study area according to characteristics of the natural environment and relative location, which should reflect differential opportunity levels for economic and demographic change. Individual community membership of clusters would no doubt change if the analysis were run for different dates or cluster algorithms, but the basic spatial pattern should be stable for the time-scale considered here. There are, however, important features of the local (intra-community) environment that did change over the 20-year period, and these must be examined before testing the polarisation and diversification hypotheses.

THE LOCAL SOCIAL ENVIRONMENT

In considering any system, an important (and subjective) analytical problem is to separate the system itself from the environment in which it is embedded – particularly in the case of highly open systems, such as place-based rural communities. Certain important qualities of a place could be classified either way. Coombes and Raybould (2001) have shown that key factors in classifying non-metropolitan settlements are the degree of urbanisation, as indicated by the population size of the largest town; and concentration or nucleation of settlement, measured by the proportion of the defined community resident in urban centres. Along with remoteness and local rural population density, they have also been shown by the present writers (Smailes et al., 2002a) to be important predictors of a range of social and demographic aspects of rural communities. They represent important qualities of the place into which prospective in-migrants may move, or which out-migrants leave behind; migration itself will alter these qualities in a positive feedback loop. The size of the largest town reflects the range and kinds of urban amenities (including housing) it provides; the degree of concentration of the population into towns reflects the nature of the settlement pattern and perceptions of rurality/urbanity; and the density of the rural population in the district reflects the nature of the surrounding economic and cultural landscape. Table 5.2 illustrates the differential changes in these key qualities over the two decades in the five clusters. Because of the influence of a small number of extreme cases, medians are preferred to means in expressing typical values of the variables.

Table 5.2 Change in urban size, urban concentration and rural density 1981–2001, by cluster

Cluster	Population of main community centre			Percentage of population living in towns			Density of occupied rural dwellings (per 100 km²)		
	Median	Median	Change	Median	Median	Change	Median	Median	Change
	1981	2001	1981–2001	1981	2001	1981–2001	1981	2001	1981–2001
			(%)			(% points)			(%)
1 Tablelands/ranges	1614	1661	2.9	58.3	59.8	1.5	25	31	24.0
2 High access	2667	3454	29.5	64.2	68.4	4.2	97	169	74.2
3 Mixed farm	2044	2842	39.0	65.7	67.6	1.9	58	78	34.5
4 Coastal	2172	3040	40.0	66.6	74.8	8.2	161	231	43.5
5 Inland/remote	951	953	0.2	54.3	58.3	4.0	14	13	–7.1

Source: Calculated by the authors from Australian Bureau of Statistics 1981 and 2001 Censuses (ABS & Space-Time Research, 1990; ABS & MapInfo, 2002); areas calculated by GIS

Over the 20 years, a growing disparity in the local social environments emerged between the two spatially most extensive clusters 1 (tablelands/ranges) and 5 (inland/remote) on the one hand, and the more accessible and coastally located clusters 2–4 on the other. The median size of the main towns in the former clusters is both much smaller than in the latter clusters, and has shown negligible change, while median size rose substantially in the high-access, mixed farm and coastal clusters 2–4. The median percentage of community population living in urban settlements (including subsidiary townships) increased over the two decades in all clusters, most significantly in the small, mainly coastal communities of cluster 4, in which the median value of urban concentration reached 75 per cent by 2001. Clusters 1 and 5 are again distinguished from the others by their very low densities. The median density of rural settlement showed an actual decline only in cluster 5 (inland/remote), thereby helping to account for its increase in urban concentration; cluster 1 (tablelands/ranges) showed a small gain. By contrast, the other three clusters have all shown very significant gains in their already substantially higher median rural settlement densities. Clearly, demographic growth in these clusters was not being fuelled by town growth alone.

Over the 20-year period, substantial changes occurred in the distribution of population within the studied communities, as between the central town, rural and minor town[2] components. Space does not allow for detailed tabulation, but the main community centres grew (in aggregate) in all five clusters, though with a clear contrast between low growth of 5–7 per cent in the two most rural clusters (inland/remote and tablelands/ranges), and sharp rises of 21–40 per cent in the other three clusters. As to the 'rural' component, despite the farm crisis it actually fell only in the rural/remote cluster, where a net decline of some 19 000 persons (12 per cent) occurred. The greatest increases were in the high-access and coastal clusters, with rises of 34 and 22 per cent respectively.

The most dramatic changes, however, occurred in the minor clustered settlements – even the inland/remote cluster registered a modest 13 per cent aggregate increase, and the tablelands/ranges 77 per cent. In the three more closely settled clusters (2–4), minor town increases ranged from 125 to 138 per cent. In 1981, minor centres were found in 110 of the communities in clusters 2–4, and 88 of these cases registered increases – in some cases, massive increases – by 2001. In cluster 2 (high-access), almost all increased. Additionally, new minor centres emerged in a further 16 communities in clusters 2–4, many of them on the coast. Even in the two more sparsely settled clusters, growth had occurred in minor settlements in almost half of the communities where they were present in 1981. The tablelands and ranges (cluster 1) experienced localised but substantial growth, and by 2001 new small centres had appeared in a further 14 of its 110 communities. Overall, this change in the settlement pattern takes a striking and spatially consistent form, with growth occurring particularly along the coast, or close to the metropolitan areas and the larger country towns.

To summarise, overall population trends show a decline in cluster 5, particularly its rural component; small gains in cluster 1; and sharp increases in the other three clusters, with the greatest absolute increases in the main community centres, but the greatest relative increases in minor clustered settlements.

EXAMINING THE HYPOTHESES

POLARISATION

We turn now to examine changes over the two decades that are much more dependent on decision-making by residents (including recent or prospective in-migrants) of the local communities themselves, and can be influenced by the mobilisation of local human and social capital. Have the differential opportunities presented by the contrasting environmental endowments of the five clusters been marked by increasing polarisation or divergence? To answer this question, we compared the extent of change in the five different clusters, using two groups of key indicators – demographic and economic. Mean percentage changes in each indicator between 1981 and 2001 were calculated for the communities in each cluster.[3]

Taking the demographic indicators first, we found surprisingly little variation between the cluster averages in several important aspects of population structure and change. By 2001, the proportion of children aged 0–14 years had declined to a very similar level of around 22–23 per cent in all clusters, and by a similar percentage change of 4–5 per cent. The ageing of the potential workforce was very substantial in all clusters, with the older age groups (40–64) rising from 40–42 per cent of the working-age population in 1981 to 53–54 per cent in 2001. The mean proportion in the retirement age groups also increased in all clusters, but by different percentages. It rose by 6.1 per cent in the inland/remote cluster, but only by 3.0 per cent in the high-access cluster – in effect reducing the inter-cluster contrasts to a similar level of around 14–16 per cent aged over 64.

The other three demographic measures (mean changes in fertility, masculinity and residential mobility) had all declined in every cluster, but these declines have been more uneven, and show more evidence of diverging trends between the clusters. In each case, the major contrast that emerges is between clusters 5 (inland/remote) and 4 (coastal). Traditionally, rural areas have had high fertility offsetting net out-migration, and a high ratio of males to females, due to the male-domi-nated nature of the farm workforce and shortage of jobs for females. Residential mobility, as indicated by the proportion of people who have changed address in the five years preceding the Census, is typically low in rural areas. Although the mobility data include people moving house within the same town, they provide a surrogate in-migration measure, since they can only be recorded at a person's new address.

The inland/remote cluster had retained all these traditional rural characteristics to the greatest extent, with the lowest mean percentage changes (–1.4, –1.7 and –0.7 in fertility, masculinity and residential mobility respectively). Despite these minor declines, the cluster retained the highest mean fertility ratio of 40.6,[4] highest masculinity ratio (107.5) and lowest residential mobility (35.1). By comparison, the coastal cluster had the greatest changes (–6.6, –5.4 and –4.0) in the same three variables. Fertility by 2001 had dropped to only 33.3, and masculinity to 98.0, so that females actually outnumbered males. Despite a fall of 4 per cent since the 1981 Census (which was taken at the peak of the counterurbanisation trend), the coastal cluster had the highest average residential mobility.

Summing up, there is evidence of divergence between the persis-tence of traditional rural patterns in cluster 5 (rural/remote) and the trend away from such patterns in cluster 4 (coastal) in three of the six indicators. This divergence does not constitute polarisation, for the other three clusters take various intermediate positions, in no consis-tent order.

Six further indicators were chosen to examine the economic evidence for polarisation. The first three relate to mean changes (1981–2001) in composition of the workforce by industry; these indi-cators are primary industry employment by number and by per cent of the total workforce, and an index of industrial diversity. Over the 20-year period, the study area in aggregate lost almost 42 000 agricultural jobs, or just over 20 per cent of the 1981 workforce, with all five clusters sharing this percentage loss. These changes produced very similar results in clusters 2 (high access) and 4 (coastal) where by 2001 farming employment had shrunk by 14–15 per cent, to employ a mere 11–12 per cent of the workforce. By contrast the two most traditionally rural clusters 1 (tablelands/ranges) and 5 (inland/remote), despite losing well over 20 per cent of their respective farm workforces, still depended heavily on farming, which accounted for a quarter of the 2001 workforce in cluster 1 and over one-third in cluster 5. Cluster 3 (mixed farming) occupied an intermediate position, with 19 per cent still employed in agriculture. When interpreting these figures, an important caveat is that the Census asks only for a person's main occupation, and does not record pluriactive part-time workers in farm households, who may be very important in clusters 2–4 where there are many more small farms and access to alternative urban employment is easier. Off-farm income has great importance for farm survival (Barr, 2002) and the Census employment data are influenced by self-perception of farm residents in completing the form.

These changes in farm employment obviously have an impact on the diversity of employment available in rural areas. This is measured by the Gini coefficient, using 13 Census workforce employment categories (not including the unemployed). The index may vary from 100 (employment shared equally between all industries) to zero (all employment is concentrated in just one industry). In practice, the values in our study range from 16.0 to 69.0. Clearly, community sustainability is favoured by access to a diversity of employment, pluriactivity, availability of part-time work, and non-dependence on a single large, potentially vulnerable industry. Industrial diversity is an important but complex variable, with change in any one industry producing time-lagged derived and induced effects on others. In a relatively evenly balanced economy, heavy loss of jobs in one industry (for example, farming) will normally reduce industrial diversity, and this has occurred in clusters 1–4, where the index fell by between 2.5 and 5.6 per cent, with the greatest fall in the coastal cluster.[5] In cluster 5, however, farming was formerly so heavily dominant that its shrinkage has produced an actual slight rise (+0.3 per cent) in the index.

The other three economic indicators used are changes in the size of the workforce, percentage workforce participation by people aged 15–64, and per cent unemployed, respectively. The three closely settled clusters 2–4 have all experienced large (30–40 per cent) gains in the mean workforce size, but more varied changes in participation rates and unemployment. As with the demographic variables, the main evidence of diverging trends is the contrast between clusters 4 and 5. The inland/remote cluster has the highest workforce participation rate, the lowest unemployment, and only a small increase in unemployment over the two decades. Jobs outside of farming are scarce, and most people displaced from farming have migrated out of the community rather than remain at home unemployed. The communities in the more attractive environment of the coastal cluster 4, on the other hand, have on average the lowest participation rate, highest unemployment and most rapid rise in the proportion unemployed.

To derive summary measures of the extent to which change over the 20 years produced convergence or divergence among the clusters, the 1981–2001 percentage change in the key variables was calculated for each community, and the median percentage change obtained for each cluster (see table 5.3). In most variables, the median for the inland/remote cluster (5) was at or close to one extreme, followed by the tablelands and ranges cluster (1). The coastal cluster (4) was usually at or near the opposite extreme, with the high-access cluster (2) close to it. The remaining mixed farming cluster (3) was variable, but on average occupied the middle ground. If the clusters are arranged in this sequence, the extent to which each variable conforms to or diverges from this sequence can be seen, and each cluster given an average ranking for the 15 variables combined (see table 5.3).[6]

Clearly, the ranking of the clusters on this table is no more than an expression of broad trends, and while the change in some variables showed huge differences between ranks 1 and 5, in others the differences were quite small. The mean rank was not weighted to take account of these differences. Nonetheless, the average ranks do place the clusters in the suggested sequence, and this emerges particularly strongly in the case of the important total and urban population variables. Apart from the population variables, though, the changes do not appear to be moving towards a simple dichotomy between the clusters, or even a continuum with a shrinking middle ground; each cluster retained its own distinctive combination, or trajectory of change.

From the evidence above, we conclude that there is no simple polarisation between two dichotomous groups. There is certainly divergence at the extremes; a strong contrast exists in many of the key variables between clusters 5 (inland/remote) and 4 (coastal), which thus lie at opposite ends of the proposed 'depopulating and declining inland/growing and prosperous coastlands' opposition. Closest to the inland/remote cluster, and retaining many aspects of the traditional rural profile, is cluster 1

Table 5.3 Clusters ranked by median percentage change on 15 key indicators, 1981–2001

Variable	Median percentage change 1981–2001					Ranking of clusters by degree of change				
	Coastal (Cluster 4)	High access (Cluster 2)	Mixed farm (Cluster 3)	Tablelands/ ranges (Cluster 1)	Inland/ remote (Cluster 5)	Coastal (Cluster 4)	High access (Cluster 2)	Mixed/ farm (Cluster 3)	Tablelands/ ranges (Cluster 1)	Inland/ remote (Cluster 5)
Percentage of population aged <15 years	–4.38	–4.30	–5.01	–4.82	–4.96	2	1	5	3	4
Percentage of population aged >64 years	5.03	2.95	5.10	4.89	6.07	3	1	4	2	5
Children aged 0–4 as percentage of females aged 15–44	–6.51	–5.21	–3.70	–3.22	–1.48	1	2	3	4	5
Age 40–64 as percentage of age groups 15–64	11.81	12.18	11.31	12.52	12.31	2	3	1	5	4
Males per 100 females	–3.94	–1.83	–2.64	–2.33	–1.88	1	5	2	3	4
Percentage changed address last five years	–3.37	–1.79	–0.73	–1.44	–0.40	1	2	4	3	5
Percentage employed in primary industry	–7.13	–5.20	–7.26	–4.64	–5.63	2	4	1	5	3
Number employed in primary industry	–21.07	–19.63	–17.77	–23.30	–27.86	3	2	1	4	5
Industrial diversity index	–5.95	–4.01	–2.39	–2.68	0.17	1	2	4	3	5
Size of workforce	38.44	40.27	19.06	–4.89	–14.31	2	1	3	4	5
Percentage of ages 15–64 in workforce	–0.86	0.64	0.82	–1.96	–1.22	3	2	1	5	4
Percentage of workforce unemployed	2.60	1.31	0.47	1.19	0.32	1	2	4	3	5
Total rural population	23.55	35.38	10.19	–3.93	–18.81	2	1	3	4	5
Total urban population	39.06	32.24	17.87	–2.35	–6.69	1	2	3	4	5
Total population of community	38.64	35.78	17.32	–2.17	–12.16	1	2	3	4	5
Mean rank						1.7	2.1	2.8	3.7	4.6

Source: Calculated by authors from Australian Bureau of Statistics 1981 and 2001 Censuses

(tablelands and ranges), while the other two clusters – 2 (high access) and 3 (mixed farming) – each have distinctive trajectories of change, and on most variables lie somewhere between clusters 1 and 4.

INCREASING HETEROGENEITY

Heterogeneity among rural communities may arise not only between, but also within the defined clusters. Although the changes over 20 years in the cluster means have been relatively muted, many individual communities may change their positions within those distributions. After all, the clusters are based on qualities of the natural environment, relative location and potential farm productivity, and take no account of the differential endowment of human, social and institutional capital. To test whether rural communities are becoming more heterogenous, we sought to determine the degree of variation about the means of the key variables within each cluster, and whether this variance increased over time.

Table 5.4 compares the standard deviation for 1981 with that of 2001 for 14 selected variables.[7] A fall in the standard deviation over time (ratio less than unity) indicates that communities are becoming more tightly clustered about the mean, while an increase (ratio greater than unity) indicates a greater scatter and diversity. A ratio of exactly 1.00 shows that no change in variance occurred.

Counter-intuitively, the results showed that, in the majority of the 70 cases tabulated, the communities within each cluster tended to become more similar rather than more diverse. While many of the changes were not large enough to be statistically significant, the general tendencies are unmistakeable. However, there were striking differences between the clusters, cluster 1 (tablelands and ranges) being the only one where the constituent communities became more diverse in a majority of key variables. By contrast, those in clusters 2, 3 and 4 became more similar in most respects. In particular, cluster 4 (coastal) showed a lower variance on every one of the key variables tabulated. Trends in cluster 5 (inland/remote) were much more varied.

An important reason for changes in diversity within the clusters is likely to be the fall in the relative importance of agriculture as the fundamental base for rural settlement, and the rise of amenity and accessibility considerations both in attracting a new adventitious population and in stemming out-migration. In the coastal cluster, the uniform reduction in diversity almost certainly arises from the attraction of the coast itself and its scenic backdrop, bringing in new migrants with characteristics sufficiently similar to blur the formerly greater differences between communities. Along the whole New South Wales and part of the Victorian coast, either the community centre towns or minor urban clusters – or both – grew substantially over the study period, with in-migration a

Table 5.4 Changes in variance within each cluster for 14 selected variables, 1981–2001 (standard deviation 2001 as ratio of standard deviation 1981)

Variable	Cluster 1 Tablelands/ ranges	Cluster 2 High access	Cluster 3 Mixed farm	Cluster 4 Coastal	Cluster 5 Inland/ remote
Percentage of population aged <15 years	**1.30**	0.89	0.85	0.80	**1.06**
Percentage of population aged >64 years	**1.10**	1.00	**1.08**	0.88	0.96
Children aged 0–4 as percentage of females aged 15–44	1.00	0.71	0.72	0.84	1.00
Age 40–64 as percentage of ages 15–64	**1.10**	**1.01**	0.87	0.70	**1.18**
Males per 100 females	**1.06**	0.67	**1.16**	0.62	0.99
Percentage changed address in last five years	**1.09**	0.92	0.82	0.86	**1.12**
Percentage employed in primary industry	**1.03**	0.80	0.86	0.82	1.00
Industrial diversity index	**1.12**	0.90	0.75	0.69	**1.16**
Size of workforce	0.99	0.85	0.95	0.86	0.98
Percentage in workforce (ages 15–64)	**1.04**	**1.16**	0.98	0.86	**1.17**
Percentage of workforce unemployed	**1.02**	**1.14**	0.99	0.73	0.98
Total rural population	**1.01**	0.92	0.87	0.86	0.96
Total urban population	0.98	0.87	0.98	0.94	0.98
Total population of community	0.97	0.86	0.97	0.87	0.98
Number of increases in variance	**10**	**3**	**2**	**0**	**5**
Number with variance unchanged	1	1	0	0	2
Number of falls in variance	3	10	12	14	7

Source: Calculated by authors from Australian Bureau of Statistics 1981 and 2001 Censuses

major cause. The increase in diversity in clusters 1 (tablelands and ranges) and 5 (inland/remote) relate to the more restricted distribution of perceived amenity, and the more selective growth/decline patterns among the communities and their central towns.

Though not reported in detail here, our preliminary investigations of the impact of amenity suggest that strong correlations exist between this variable and many of the key indicators discussed in this chapter, using the 1981 distribution of employment in the accommodation, entertainment, cultural and personal services industrial categories as a proxy for amenity. These activities are ubiquitous in all communities, but should be proportionately higher in places attracting many tourists and visitors. Assuming that the holiday travel behaviour of Australians is based on their perception of 'amenity', the variable proved a crude but effective surrogate. Strong positive correlations were observed between this surrogate and residential mobility, the proportion of retirement age groups and overall population growth from 1981 to 2001 – particularly growth in minor townships and the rural population element. There were strong negative relationships with, for example, workforce participation and change in the industrial diversity index. It should be noted that these correlations were strongest in the coastal cluster, but very weak in the high-access cluster, where amenity is far less important than accessibility as an arbiter of change.

CONCLUSION

This chapter has shown that the widespread belief that a growing demographic, social and economic divide is driving coastal and inland regions of Australia further apart fails to represent the diversity of rural Australia's experiences over the past two decades. While it is certainly true that the eastern coastal belt is becoming increasingly unlike the remote inland regions of the country, which have long carried the nation's self-image, the remaining rural communities do not fall conveniently into either of these polarised categories. Rather, the three distinct clusters that they form each have their own particular distinguishing socio-demographic and economic qualities, so they occupy the middle ground in different sequences for different variables. The widespread impression of an emerging dichotomy, or a simple continuum, seems partly justified when one looks at the changes in certain variables, such as total population or population of urban centres – both very frequently used as shorthand measures of success or failure in rural communities. It is important to realise, however, that 'town' should not be equated with 'community', for each community also contains a catchment area whose demographic characteristics make a large and varied difference to the overall community composition.

Reflecting on this study, it may be asked why the entire argument was based on a set of clusters defined on the basis of environment, spatial location and settlement history in an age when environmental determinism is discredited and country town economies are said to be 'uncoupling' from their former close links with agriculture (Stayner &

Reeve, 1990). The answer lies in the long history of productivism and the powerful forces of inertia, established institutions and invested capital, as well as the many intangibles that drive rural residents to strive for survival of their own communities.

The ten clustering variables we used were chosen to reflect the potential capacity of a community based on land-based industry to sustain itself over time. They are all forms of natural and locational capital that accrue to the community by virtue of its location in the broader environment and its competitive placing within national and regional space, rather than being specific qualities of the place itself. They are likely to change relatively slowly over time, and should therefore produce a relatively stable typology. We would argue that the emergence of tendencies towards a post-productive transition in Australia is unlikely to produce a radically new classification of communities by locational types, for the following reasons.

Agriculture and extractive industry have been the main motivation for rural settlement in Australia for much the greater part of its European history, and the enduring character of the national space economy evolved under these productivist conditions.

When, very much later, post-productive or consumptionist motivation began to influence the occupance of rural space, it was guided by many of the same environmental and locational considerations, though for different reasons. In introducing a new set of demographic trends, it thus reinforced rather than substantially changed the space economy. Many elements of natural capital that had once attracted farming settlers now attracted alternative lifestylers, early retirees, urban refugees, developers and others motivated by the consumption of perceived amenity.

At the social catchment or local community scale of resolution used in this study, the spatial units are large enough for productionist and consumptionist motivations to operate in parallel on different types of site, particularly after zoning regulations began to protect the best agricultural land.

The data collected for the cluster analysis were based on the 1981 situation. By then, most rural communities had already been through the weeding and sorting processes of at least a century of boom and slump cycles, occupation of the ecumene was practically complete, and it appears unlikely that the clusters would have been substantially different using, say, 1971 or 1991 data instead.

The findings of relative stability in the defined clusters, and the relatively limited changes in the means and standard deviations of the statistical distributions in each cluster, do not of course mean that no important changes are occurring, for individual communities can shift their relative positions in the distribution quite radically with no change to the mean or variance. Indeed we would be amazed if it were not so, because the clustering variables we have considered reflect only natural and locational attributes, and do not allow for the very different resources that communities may have of produced, institutional, human and social capitals, or the skill with which community leaders harness and co-ordinate these resources.

On top of the cluster analysis variables, we have identified a number of factors that express qualities of the internal local environment within each community: main town population size, urban concentration, density of rural population in the social catchment, and 'amenity' – a non-productivist factor that, together with accessibility/remoteness, is becoming an increasingly important driver of rural community change. All of these can change appreciably over two decades, in reciprocal relationship with demographic change. All of them affect sustainability, and deserve more attention than the scope of this chapter allows. Although these factors may reflect something of the produced and institutional capital likely to be present in a community, they still say nothing of the human and social capital available, nor of the cultural and sociological qualities of life in the community, such as class and faction structure, openness to newcomers, integration and leadership. The case studies in the research team's earlier volume (Cocklin & Alston, 2003) provide ample evidence of the importance of these qualities, and the huge difference they can make to the quality of life. All of these factors are likely to make for increased local variation in the trajectories of communities similarly endowed with natural, locational and produced capital.

Moreover, in a climate of neoliberalism where responsibility for regional development is being devolved downwards, and local communities are expected to pull themselves up by their own bootstraps, the traditional competition between communities for resources and new ventures is likely to be intensified unless carefully managed. Elsewhere, the point has been made that sustainability needs to be sought at a level somewhere between the 'region' – a large and somewhat artificial construct, in theory possessing an adequate critical mass to achieve scale economies – and the local community, which possesses the necessary cohesion, social capital and group identity, but in most cases lacks scale and critical mass (Smailes & Hugo, 2003: 101). This can best be achieved through coalitions of collaborating groups of communities.

Despite the great improvements in communications over the last 20 years, the very low densities of cluster 5 communities make it difficult to achieve the level of interaction needed to share resources and co-operate in effective groups. It should be realised, however, that in different environmental conditions, sustainability can be achieved at different levels of population size; numbers can be stabilised at a new lower level, and nothing is achieved by seeking some magic minimum population level for community viability to apply across the board, and particularly not by using town size alone as a crude proxy measure.

This potential for change caused by communities jockeying for position within the established environmental clusters raises the question of whether a new cluster analysis based on the outcomes of change over the 20 years would yield different results to the present one, which is based on initial factor endowment – and if so, what relation the geography of the one set of clusters would bear to the other. This question must be deferred for further investigation, along with the relationship of the clustering pattern discussed here to those of Barr (2000), who recognises 12 clusters across the whole nation, based on a large but exclusively agricultural data base, and Holmes (2004), who recognises six occupance types based on the relative precedence of production, consumption and protection (conservation) as the basic motives for occupation of rural land. The problem of achieving sustainability for the system of rural communities that occupy our national ecumene, including resilience and the ability to absorb inevitable change, is a huge one that requires identification, classification and regionalisation of rural communities so as to develop policy differentiated according to regional needs. We see this chapter as a contribution to this end.

ACKNOWLEDGMENTS
The writers acknowledge their debt to David Gerner, Justin Nottage and Geraldine Mason for their invaluable assistance with the GIS and database preparation at varying stages in the writing of this work, and the ARC for the funding which made it possible.

NOTES

1   The eight ratio scale input variables were: median annual rainfall, median altitude, median slope, soil fertility, agricultural productivity, remoteness, gross regional population density, and duration of European settlement. They were transformed where necessary to give a normal statistical distribution, and scaled to give values ranging between +1 and −1. The two binary presence/absence variables (coast/inland and irrigated/dry) could not be used in the clustering procedure, but are important in interpreting the results.

2   It should be remembered that small agglomerations of less than 200 people, including holiday shack colonies, rural hamlets, isolated new subdivisions etc. are, in most cases, not separately identifiable from the Census data, and are here included as 'rural', as well as dispersed homesteads. Those which are separately identifiable are treated as minor centres within the relevant community.

3   These variables have a close to normal statistical distribution, so that there is little to choose between mean and median as measures of central tendency.

4   Children aged 0–4 per 100 females aged 15–44.

5   In 1981, the 'not elsewhere classified' category was very large (8.5 per cent of all employment). By 2001, these respondents being much better distributed among the other categories, it (together with 'industry not stated') was only 2.3 per cent. Part of the decline in the index is thus a statistical artifact.

6   The median values for percentage change, used in Table 5.3 to avoid the influence of extreme values, differ slightly from the equivalent *mean* values for change discussed earlier in the chapter.

7   The values for each community, on each variable, were expressed as z-scores based on the mean for all 414 communities, and transformed to ensure a normal statistical distribution, before calculating the SD for the clusters.

PART II

ISSUES IN CONTEMPORARY RURAL AUSTRALIA

6

GLOBALISATION, AGRICULTURAL PRODUCTION SYSTEMS AND RURAL RESTRUCTURING

Geoffrey Lawrence

A number of global processes have fostered the restructuring of agriculture, and of rural regions, in Australia. These include the progressive transnationalisation of capitalism, the increased global mobility of capital and people, the redirection of the activities of the nation state, and the greening of western societies. Farming is being transformed as competitive pressures are exerted by international markets, as consumers demand 'safe' foods and as an awareness develops of the serious environmental consequences of past and present agricultural practices. The penetration of farming structures by corporate capital has been hastened by the deregulation of industry. Some rural regions are experiencing contraction linked not necessarily or directly to agricultural change, but to the policies of business and the state as they withdraw services and reduce infrastructural investment. Other regions are improving their economic performance based upon opportunities in industries such as tourism, recreation and leisure – that is, as part of a 'post-productionist' future. The 'unevenness' resulting from these changes is viewed as an effect of capital accumulation under conditions of neoliberalism, with the latter policy orientation placing great faith in both market forces and individual/community-based actions to achieve positive outcomes. This chapter[1] traces recent agrifood – and broader rural – restructuring in Australia, arguing that many of the changes occurring are not conducive to the retention of natural capital, or to the building of social capital, in the regions.

GLOBALISATION AND AGRICULTURE

Globalisation is a process through which space and time are compressed by technology, information flows, trade, and power relations, allowing distant actions to have increased significance at the local level (Gray & Lawrence, 2001a; McMichael, 2000; Wiseman, 1998). There is an obvious economic aspect to this, with commercial firms – particularly those in the corporate and finance sectors – organising activities across the boundaries of nation states in order to maximise profits. Another aspect is that of regulation. While the democratic nation state's legitimacy has been achieved through the incorporation of the wishes of its citizens into nationally applied policies, processes of globalisation tend to undermine nation-specific laws and statutes. Indeed, the latter are often interpreted as restrictions on trade, with bodies such as the World Trade Organisation creating a powerful legal framework to ensure that nation states conform to a globalisation 'agenda'.

The above assessment notwithstanding, it is important to recognise that the process of globalisation is contradictory, incomplete, complex and contested (see Almas & Lawrence, 2003; Gray & Lawrence, 2001a; Pritchard & Burch, 2003). A discourse of progress-through-change is embraced by those benefiting from globalisation. A new world order is considered to be emerging, one in which free trade, increased interaction, better global/local integration, and the sharing of a genuinely global culture will be the basis of strong economic opportunity for those nations abiding by the (global) rules. A more critical assessment – such as that provided by geographer Doreen Massey (1994) – reveals that the 'power geometry' of globalisation reinforces existing structures and processes of advantage and privilege at the same time as it marginalises and excludes those with limited economic power, or who are unable to provide the ingredients for extended global capital accumulation. Like other, more recent, assessments (Gray & Lawrence, 2001a; Marsden, 2003), Massey identifies globalisation as a differentiating rather than a homogenising force. Findings from contemporary studies (Appadurai, 1990; Giroux, 2002; Gray & Lawrence, 2001a; Herbert-Cheshire, 2001; Lash & Urry, 1994; McMichael, 2000) have indicated that globalisation, in conjunction with neoliberal policy orientations:

• includes transnational practices that are largely independent of any nation state and that involve 'flows' of money, ideas, people and information, as well as 'environmental flows'; for example of water, oil and wastes

- heralds, via labour market deregulation, the 'disciplining' of the working class as union strength is weakened by the 'untamed' nature of capital movement
- is catalysed by the western world and finds expression, but also produces resistance, in localised sites throughout the world
- has begun to produce 'global' citizens whose identity is based not upon the 'nation' and its history, but upon their status within wider networks of interaction
- has called into question rights, obligations and traditional laws in relation to such things as trade, bi-lateral negotiations, and local social engagement
- has encouraged the development of 'societies' that are no longer coterminus with the nation state and that form new socio-spatial groupings through-out the world (the 'green' movement is an example)
- has fostered a broad cosmopolitan democracy defined by overlapping networks of power
- offers human society an ideology of 'artificially conditioned optimism' (Beck, 2000)
- has, through the rhetoric of neoliberalism, sought to encourage individuals (and the 'community') to take responsibility for their own problems

While the reality of capitalist expansion includes the spectacular failures of Enron, Walmart and Xerox in the United States, or HIH or AMP in Australia, as well as the financial successes of Nike or Microsoft (see Giroux, 2002: 152), there is a certain inescapable logic to the growth and spread of corporate entities. The corporate sector views locations throughout the world as differentiated sites for potential investment. Desiring such investment, national governments are reluctant to impose new taxes, constraints or pro-labour policies in the face of the loss of possible new investment and/or the withdrawal of current investment. What has occurred, instead, has been a deregulatory policy thrust, backed by organisations such as the World Trade Organisation, and viewed by firms as being crucial to their economic success. In order to compete globally, firms must adopt new sophisticated and productivity-generating technologies, minimise labour costs and reduce their tax liabilities (Ross & Trachte, 1990). Each of these actions has important impacts on the ways social and economic life is organised at the local level. This is especially so for agricultural regions.

Australia's agricultural economy has had an international focus for over a century and a half, with the shipping of commodities such as wool, wheat, lamb and beef to markets in Europe and, later, Japan and the United States. Despite the economic importance of agriculture, cost-price pressures and government policies since the 1970s have encouraged the process of 'adjustment', with producers on many medium-sized holdings either leaving agriculture or increasing the scale of production through land acquisition (Gleeson & Topp, 1997; Hooper et al., 2002). It would therefore be wrong to believe that agricultural structures have ossified, or that change is only a recent feature, as Davison confirms in chapter 3. Agriculture is, and has been, a dynamic industry in Australia, and producers and their representatives have been acutely aware of factors affecting supply and demand. Farmers have readily adopted advanced chemical, mechanical and biological technologies developed by the corporate sector, have welcomed extension advice provided by the state, and have relied upon the state and banking sectors to assist in the financing of new equipment and other purchases (Burch et al., 1996; Lawrence, 1987). Just as in the United States, Australian farms – to ensure viability – are becoming larger in size and more productive, with some 30 per cent of broadacre farms producing 70 per cent of the industry total (Robertson, 1997: 84). Farming is also becoming more industrialised (Burch & Rickson, 2001; Napier, 1997). The result has been continuation of the 'productivist' model of agriculture that has also come to characterise farming in the United States, Canada and many European nations (see Buttel, 1994, and Commins, 1990; and for Australia, Gray & Lawrence, 2001a; Lawrence, 1995).

The argument has been advanced that the present trajectory of farming in Australia is quite different from the past: farmers who were once (relatively) independent commodity producers, selling products under state-authorised marketing arrangements, are becoming progressively integrated into the industrial food sector (Burch & Rickson, 2001). Socioeconomic relations are changing as control over the production process moves off-farm, with entities such as banks, food processors, and supermarkets having a greater say in production. These new relations with the farming sector offer transnational corporations (TNCs) greater profit-making opportunities (Burch, forthcoming; Burch & Rickson, 2001), while state deregulation – in an era of rapid advances in technology, increased volatility in global markets, and the concentration of food processing (see Napier, 1997) – has ensured that the future of 'family farming' has become inseparable from the aspirations of firms in the corporate sector.

THE FAMILY FARM AND CORPORATE AGRIBUSINESS

PRODUCTION CONTRACTS

In many countries, the reorganisation of agriculture by various fractions of capital has been accompanied by a shift towards contract production (Moreira, 1998; Wolf & Wood, 1997). Although the bulk of Australia's agricultural production does *not* occur under contract, some industries are showing a significant move towards contract production and associated changes in managerial authority. Indicative of this trend is the transformation in the potato industry in Tasmania, from production for the fresh-food market to the current situation where some 95 per cent of the crop is grown under contract for the frozen food market, dominated by food processors such as JR Simplot (ex-Edgell-Birds Eye) and McCain. The number of producers has also declined steeply, from 7000 potato growers 40 years ago to 550 in the mid-1990s (Fulton & Clark, 1996) and 330 today (Burch & Rickson, 2001: 174). The surviving farmers have entered into production contracts that specify volume, variety and price. According to Miller (1996), most growers favour contracts because of the economic benefits of linking with 'globally-oriented' food corporations. However, contracts are also a cause of concern for farmers, because of the loss of managerial autonomy (Rickson & Burch, 1996). In an era of declining state involvement in extension (Vanclay & Lawrence, 1995), farmers increasingly rely on agronomic advice and crop management options provided by the processing companies. Company field officers and input supply companies' representatives generally recommend the purchase of greater volumes of agri-chemical and other inputs, placing the producers in a dependent relationship to agribusiness input industries and – as a result of borrowing for such purchases – to finance capital (Fulton & Clark, 1996). Thus, while potato producers still own their farms, they exhibit high levels of capital penetration in the form of market, technology and financial dependence.

Tariff reductions since the 1980s, which have allowed more imports into Australia (as part of global sourcing), have resulted in prices being negotiated downward (Fulton & Clark, 1996). A general concern expressed about global sourcing is that, if Australian farmers become uncooperative, a company may contract with farmers in alternative, off-shore sites to fill the company's needs (Burch, forthcoming). This adds an external element of control – particularly in the negotiation of product price. The company gains flexibility, while the individual farmer, who is locked into production of commodities specifically for one contracting company, loses negotiating power (Burch et al., 1992). Despite the apparent weakness of the position of contract growers, a blockade of the McCain's potato factory in May 2001 – and the threatened extension of the blockade to Simplot – demonstrated that desperate farmers will be prepared to run the risk of losing contracts. In the event, both processors agreed to the growers' (modest) demands for a $30 a tonne price increase (Courtney, 2001). Contract production also occurs in relation to other agricultural commodities, such as processing tomatoes (Pritchard & Burch, 2003) and chicken meat (Dixon, 2002). The contract production relations that typify the production of vegetables and other processed crops are increasingly evident in some of Australia's most important farming areas, such as the Darling Downs region of Queensland, in northern New South Wales and in Tasmania (Burch & Rickson, 2001; Miller, 1996).

DEPENDENCE ON FINANCE CAPITAL

The rural crisis of the mid-1980s was fueled by high interest rates (above 20 per cent in several years) following deregulation of the banking industry in 1984 and by the 'market share' policies pursued by banks, which included lending to rural producers at levels beyond their ability to service loans (Smailes, 1996). Unfortunately for producers, much of their borrowing was closely followed by a slump in export prices for most agricultural products. While interest rates have now fallen to low levels (5–6 per cent), commodity prices have only marginally improved and input costs have continued to rise. As a consequence, the average return on agriculture is currently 1.05 per cent, and it is believed that, in order to increase this return to the marginally acceptable rate of 5 per cent, the prices for farm produce would need to increase by over 23 per cent – something viewed as 'unrealistic' (Higgins & Lockie, 2001a: 179).

Many farmers remain 'caught', needing to borrow for inputs but unable to cover input costs from the sale of commodities. In such circumstances, the repayment of outstanding loans has been impossible. It is estimated, for example, that some 80 per cent of Australia's broadacre farmers were unprofitable for much of the decade of the 1990s (Robertson, 1997), with the ratio of farm debt to the gross value of farm production (with inflation factored in) rising from 59 per cent in 1984–85 to 76 per cent in 1994–95 (Gleeson & Topp, 1997). As the overall level of debt has mounted, farmers have left the industry. In the ten years up to 1994–95, the number of commercial farms (with operating surpluses of more than A$18 000) fell from 130 281 to 115 368, with the decline – at a rate of over 1 per cent per annum – predicted to continue for years to come (McKenzie, 1997: 3). Debt problems have been linked to the pressure to expand farm size, with Hooper et al. (2002: 1) noting a repeated 'pattern of expansion followed by difficulty managing debt ... over the past half century'. This difficulty is particularly apparent during periods of severe drought, such as the farming sector experienced in 2002–03 (Martin et al., 2003). Argent (1996: 283) suggests that an 'essentially globalized' form of banking has become dominant in Australia since deregulation of banking in the mid-1980s. Consequently, farmer–finance company relations – built, since the Second World War, on a reasonably reliable mix of trust, security and harmony – have been replaced with a new regime where bankers blame world financial dealings (rather than their internal policies) for instability and volatility. Decisions by banks to close unviable branches have had quite profound effects on smaller country towns, removing 'vital public and private services' from those towns (Argent & Rolley, 2000: 164) and creating both anger and despair among rural townsfolk (Pritchard & McManus, 2000).

CHANGES IN THE FOOD INDUSTRY

In the past two decades, various strategies of agrifood development have paralleled the demise of state monopoly marketing boards in Australia. First, in the early 1980s, there was a merger/takeover phase in the meat, brewing and confectionary industries aimed at increasing market share in the face of a decline in the rate of domestic growth in those industries. Second, larger domestically based agrifood firms moved away from rural interests into such activities as banking, chemicals, minerals, energy and brewing. Third, companies (such as property developers and shipping firms) hitherto uninterested in agriculture or agriculture-related activities moved into food production. Fourth, production internationalised as Australian food processing firms set up branches abroad – initially in the United Kingdom or United States, but later in the Asia–Pacific region (Fagan & Webber, 1994). A fifth, more recent, strategy has been global sourcing by retailers (McMichael, 2000; Pritchard & Burch, 2003). There has also been an acceleration of corporate interest in exporting into the Asia–Pacific region – something unquestioningly desired by the Australian government (DFAT, 1994;

Pritchard, 1999) and by industry (Rogers, 1997). This new direction has been actively promoted by the Australian government through the 'Supermarket to Asia' initiative. Pritchard (1999) argues that the 'Supermarket to Asia' discourse is consonant with policies of deregulation and other neoliberal market arrangements that, while appearing to suggest that the state is reducing its involvement, actually lead to a consolidation of its role in promoting agrifood restructuring.

Although Australia wants to export high value-added goods (Garnaut, 1989), the countries to which Australia exports – many of which are 'cheap labour' nations – prefer to add value within their own territories. According to the Rural Industries Research and Development Corporation (RIRDC, 1994: 3):

> ... reliance on a strategy focussed chiefly on exporting to South East Asia will yield only limited benefit ... [These countries] have very determined and deliberate policies ... to capture value added benefits for themselves and have no intention of becoming reliant on high value added imports ... The development of their own processed food industry has been enormous over the 1970s and 1980s and their export performance outstanding while Australia has slipped.

The exporting of raw materials and semi-processed goods for final value-adding in low-cost labour areas of the Asia–Pacific region will reinforce existing problems of economic dependence, as Australia continues to rely on an agricultural sector subject to trade decline and the vagaries of climate.

RURAL RESTRUCTURING

While agricultural restructuring refers to the changes experienced on-farm as a result of pressures exerted by governments, agrifood industries and other businesses, rural restructuring is an all-embracing term that seeks to capture the changes to farming, to country towns, to regional communities and to the relationships that govern social, political and economic interaction in rural regions. In the following discussion, the changes occurring within farming, and affecting both the environment and non-metropolitan regions, are highlighted.

EFFECTS ON FARMING

Subsumption is a process through which farmers lose autonomy as their on-farm fortunes become increasingly entangled with those of agrifood and other industries (Share et al., 1991). Farming's closer links with corporate capital have been interpreted by most agricultural sociologists as representing increased subsumption (Bonanno et al., 1994; McMichael, 1994; Marsden et al., 1993). Levels of *real* subsumption (farm ownership by external capital) appear to be low, but what is important is that *formal* subsumption (farm linkages with agribusiness for credit, inputs and processing) is becoming an increasingly important feature of on-farm production. Formal subsumption is a characteristic of farming systems reliant upon productivist approaches. In the context of poor incomes from the sale of commodities, purchases of agri-chemicals and other off-farm inputs are normally funded from borrow-ings, which link farming to financial capital in a manner generally favorable to the latter. One important outcome of subsumption is, then, that the global finance system is able to make new demands on farming, with contract production being but one example.

The movement to contract farming, which Little and Watts (1994: 4) have described as one of the 'striking commonalities associated with the restructuring of agriculture' throughout the world, poses an intriguing question: since contract farming tends to provide products that are part of a newly developing section of agriculture (for example, frozen potato products, fresh vegetables and organically-produced cereal crops), are more direct links with agribusiness processing firms the key to prosperity for individual producers? In terms of the social organisation of agriculture, there is evidence to suggest that early 'gains' are masking an uncertain future (Rickson & Burch, 1996). Contract prices can be readily organised downwards when companies source globally (Burch, forthcoming), as most of the food transnationals do. However, as Burch and Rickson (2001: 176) point out, 'there remain local social, physical and cultural particularities which serve as the basis for a degree of autonomy on the part of the farmer'. For example, the regional advantages enjoyed by Tasmanian potato producers have given them 'some leverage in their negotiations with agri-food companies', explaining the success of the producers' action discussed above.

Another outcome of the restructuring of agriculture relates to what might be best termed 'detraditionalization' (Heelas et al., 1996; for agriculture see Gray, 1996; Gray & Lawrence, 2001a). Approaches to farming, the significance of family, community obligations and commit-ment to a 'rural' lifestyle have been called into question as structural changes have undermined the traditional values and lifestyle choices of farming in Australia. A pervasive individualism born of neoliberalism

(Gray & Lawrence, 2001a) is a modern-day transmogrification of farmers' belief in 'independence' – as found in measures of agrarianism (Beus & Dunlap, 1994). Other factors altering tradition include the movement of neighbours away from the industry, the need for off-farm work, the changing role of women and the general stresses of modern-day farming life (Gray, 1996; Gray et al., 1993). Importantly, a detraditionalised agriculture appears to be one in which the social capital required to defend rural social structures, and to bring about desirable change, has been eroded (see discussions in Cocklin & Alston, 2003; Lawrence et al., 1997).

ENVIRONMENTAL IMPACTS

There is ample evidence that Australian farmers believe that they are on a production 'treadmill' (see Schnaiberg, 1980, for description; and Gray, 1996, and Lawrence, 1987, for discussion). They appear to recognise that they are working harder, conforming to productivity-raising strategies (expanding their enterprises, applying agri-chemicals), and coupling on-farm production with off-farm work opportunities. They are angry that their efforts in producing food for the nation and world are being 'rewarded' with accusations of environmental vandalism. Yet they also recognise the need to incorporate new knowledges – such as options for a more sustainable agriculture – into farm and catchment planning (Lockie & Vanclay, 1997; Smailes & Hugo, 2003). What remains somewhat untheorised are the reasons for, and extent to which, farmers are prepared to over-extend natural resources to satisfy their strong desire to remain in farming.

There is now increasing recognition that productivist approaches to agriculture in Australia have caused serious, and in some cases irreversible, damage to the environment (Cocklin, in chapter 10; also AFFA, 1999; NFF & ACF, 2000). In particular, the use of agri-chemicals has been responsible for off-farm pollution of streams and rivers; overcropping and overgrazing have led to severe erosion; and tree clearing and practices such as continuous ploughing, monocropping and irrigating have resulted in widespread soil and nutrient loss and salinisation (Gray & Lawrence, 2001a; Lawrence et al., 1992). Despite the growth in voluntary community organisations such as Landcare, which bring farmers and community members together in an attempt to improve environmental management at the local level (Lockie, 2001a; Lockie & Vanclay, 1997), there is little evidence that the wider problems of land degradation are being addressed (AFFA, 1999). For example, Tonts and Black (2003: 118), in a study of the Narrogin district of Western Australia, reported that: 'Despite the activities of Landcare groups and other organisations, the problem of salinity has continued to worsen', with the area affected by salting expected to increase by as much as 117 per cent during the 15 years to 2010. Minor changes in (local) production regimes do not appear to be overcoming environmental degradation at the aggregate (catchment or national) level. In this context, it should be of concern that virtually all proposals to make Australian agriculture more sustainable involve voluntary, not compulsory, actions.

The 'environment' (plants, animals, soils, trees and water) is the major component of natural capital (Cocklin & Alston, 2003). In a nationwide study of the sustainability of country towns, researchers found that resource-dependent communities, relying upon natural capital as the basis for future growth, were increasingly vulnerable when that form of capital was being depleted in the context of the cost-price squeeze experienced by farmers (see Cocklin, in chapter 10). One of the solutions proposed to the decline of dairying in Monto, Queensland, was the development of other primary industries (goats, mining) rather than investment in the tertiary sector (or other expanding areas). The town appeared to be locked into 'resource-substitution' based upon continued land- and water-based options, which would, ultimately, increasingly diminish natural capital (Herbert-Cheshire & Lawrence, 2003). In the New England area of New South Wales, natural capital was under threat from overgrazing (low vegetative cover), with significant water quality problems arising from poor land management in both urban and rural areas (Stayner, 2003). The loss of natural capital – in the form of dryland salinity – was viewed as being the 'single biggest threat' (Tonts & Black, 2003: 131) to the achievement of sustainability in the Narrogin district. In 'Tarra', a dairying district of Victoria, land clearing, particularly for plantations, has contributed to weed infestation, erosion and the sedimentation of creeks and rivers, with declining water quality, loss of biodiversity and deteriorating visual amenity now being major concerns. Pollution and salinisation are also likely outcomes of the more intensive use of natural capital following deregulation of the dairy industry (Dibden & Cocklin, 2003). All of these studies indicated that there was an important 'local' dimension to the adoption of what are now deemed to be inappropriate land and water management regimes, and that it was unlikely that producers would be able to 'correct' the situation by acting alone (that is, without significant government funding support).

As has been demonstrated abroad (Morvaridi, 1998), contract farming does not appear to address or ameliorate damaging production techniques; rather, it seems to result in their intensification. Contracts with processors have generally left the problems of environmental damage with the farmer (Burch et al., 1992; Rickson & Burch, 1996). Short-term production goals are consistent with processing firms' strategies for global sourcing, but militate against better land and water management on individually contracted farms. When processing companies adopt the common strategy of contracting by area but harvesting by volume, it is the weight of the product – rather than the area dedicated to production – that becomes significant (Burch et al., 1992). If the contractor decides not to take the farmer's entire output, the remainder is often left to rot in the fields. This strategy, apparently uncommon in other developed countries, is highly beneficial for the contractor. The costs of *not* harvesting are borne not only by the producer but also by the environment, because such 'over-planting' increases soil loss through tillage, often results in excessive chemical spraying, and can cause nutrient depletion (Squires & Tow, 1991). Miller (1994) has also demonstrated that as farming becomes linked to food processing there is no guarantee that sustainable resource-use practices will follow.

Whether or not the apparent 'greening' of corporate food industries will lead to a reversal of this situation is far from clear at this time. However, there is evidence that some firms in the food industry are not only beginning to embrace discourses of sustainable agriculture and 'clean and green' production, but are actively seeking producers who can supply such products. Food companies are 'reaching back' into the farming sector, writing contracts with producers who can guarantee both that products will be free of artificial chemicals, and that production has occurred in a sustainable manner (Campbell, 1996; Diesendorf & Hamilton, 1997a; Lawrence et al., 1999). Contracts, in other words, can potentially be a mechanism to promote a more sustainable agriculture. Also, organically grown fresh produce is beginning to replace fruits and vegetables that are grown by conventional (chemical) methods – another indication that the food industry is listening to the demands of consumers (Burch et al., 2001). So-called 'triple bottom line' accounting is further evidence that the corporations may not be 'greenwashing' the public, and may indeed be entering a new era of 'green production' (Burch et al., 2001; Lockie et al., 2000b).

Fortunately for the future of natural capital, there appears to be a changing global consciousness to issues of the environment (Eder, 1996; Thomashow, 1995) and innovative approaches are being developed to combat such problems as dryland salinity and declining water quality (for Australia, see AFFA, 1999). In response to the demands of local and overseas consumers for more wholesome, less chemically produced, foodstuffs, Australia has sought to develop products that are 'clean and green' (DFAT, 1994; Lawrence et al., 1999). Agrifood industries hope that consumers will identify all Australian-produced foods as having these desirable qualities (DFAT, 1994), yet the dependence of farming on the chemical inputs of agribusiness – and more recently biotechnologies (Hindmarsh et al., 1998; Hindmarsh & Lawrence, 2001) – may preclude much of Australia's production from falling within this category. While 'clean and green' does not fully equate with 'organically produced', it is sobering to learn that less than 2 per cent of Australia's food is grown organically (Lyons, 1998; Monk, 1998), and that another 'clean and green' strategy – that of reduced chemical-use in production, so-called low-input farming – has not been widely adopted.

The extent to which Australia is really moving to 'clean and green' farming within an agribusiness-dominated agriculture is unclear. On the one hand, the growing links between food corporations and farmers might 'direct' producers to chemical-free and more sustainable approaches (see discussions in Allen, 1993). On the other hand, in a world of global sourcing of raw materials for the processed food industry, farmers in Australia may be exploiting natural capital in an effort to 'compete' in a sector progressively dominated by corporations that owe no loyalty to regions, their peoples or their environments.

CHANGES TO THE REGIONS

As indicated earlier, a number of writers have argued that globalisation advantages regions that are prepared to 'adjust' or restructure their activities. Lash and Urry (1994) have argued that regional space will be an important site of what they, and others (see Castells, 2000), have termed flexible specialisation. They believe that networks of small firms, with a knowledge-based workforce, will grow within the regions, allowing global capital to escape older – supposedly inflexible – structures of unionised city-based labour. Thus, as globalisation proceeds, regional localities:

... provide contexts for social interaction. And such interaction is needed to gather information, to make arrangements and coalitions, to reinforce relations of trust or implicit contract, and to develop rules of acceptable behaviour ... [Localities] enable product and process innovation to take place in relatively decentralised systems. (Lash & Urry, 1994: 284)

This positive picture for regions of advanced nations is one premised upon an increase in information flow based, in turn, upon the steady adoption of computer-based technologies. Locally-based knowledge workers (including community specialists) are expected to become the catalysts for regional development. In this way, new leaders will not only create economic opportunity but also build human and social capital, empowering individuals and communities to interact with global firms in new and productive ways.

In opposition to this sanguine view, critics have argued that 'foot-loose' corporations are likely to bypass regions that do not understand, or want to play by, the new rules of global/local engagement (McMichael & Lawrence, 2001). And what are these rules? Concomitant with an ideology of neoliberalism, they include the provision of a deregulated labour market and the relaxation of environmental restrictions so that corporate firms can invest and achieve profits without having to pay for the 'externalities' of their actions. This is the scenario of a 'race to the bottom' – one in which regions may achieve growth in 'produced capital' (that relating to growth in economic and financial benefits) while losing human and social capital, and placing natural capital under increasing stress as a direct consequence of the introduction, or continuation, of environmentally-degrading production regimes.

What evidence is available to determine what might be happening in rural Australia? According to the contributors to Pritchard and McManus' edited volume *Land of Discontent* (2000), economic policy settings have had deleterious effects on community life in rural Australia. For the contributors to Lockie and Bourke's *Rurality Bites* (2001), there is unmistakable evidence that rural communities are facing considerable problems, including land degradation and socioeconomic decline. Lockie (2001a), in the book's conclusion, argues that regional restructuring under the current policy trajectory will lead to further rural decline and to the regional dominance of corporate capital (particularly in the form of agribusiness). The building of social and natural capital is endorsed as a means of creating more sustainable local/global relations.

Pertinent to any discussion of the various 'capitals' that promote rural sustainability is the argument presented by contributors to Cocklin and Alston's *Community Sustainability in Rural Australia* (2003) that the 'stock' of human capital – the knowledge, skill and ability of individuals to make a difference – is being reduced in regional Australia, and that social capital – the networks of local engagement and the levels of social trust that provide for positive community action – has been seriously eroded in many smaller country towns. Several case studies in this book indicate that small towns must be positioned strategically within stronger regional networks if they are to survive and thrive. In this context, Bourke (2001b: 127) has argued that centres with high levels of financial, social and institutional capital (towns like Wagga Wagga, Tamworth and Bendigo) have been able to prosper in a competitive economic environment, while those with limited resources continue to suffer.

In relation to the above studies, it is useful to return to Lash and Urry's positive view of change in the regions of the advanced economies. In recognising the importance of diversifying (and 'value adding' to) their activities, some farmers are seeking to brand-name their products, often with regional labels. Others are involved in new income-generating opportunities on-farm (for example, ecotourism and farm stays) as well as off-farm (service sector employment, small business). New communication facilities – in many, though not all, regions – have provided farms and towns with direct links to the global marketplace and to global culture. Similarly, as retirement, leisure and recreation begin to figure increasingly as economic options for the regions, the farm and non-farm populations are taking advantage. Activities as diverse as wetland tours, heritage festivals or the staging of local cultural events have allowed towns to move away from their former dependence on agriculture. Such activities are not occurring evenly throughout rural space. As Dibden and Cheshire argue, in chapter 12, some locations appear to have benefited differentially. The more remote locations – unless close to a major tourist attraction – find it difficult to 'capture' such advantages. And it is these latter communities that appear to require considerable, long-term support by the state if they are to weather the storm of neoliberalism. At present, this sort of support is being denied them.

CONCLUSION

As a key process driving the capitalist economic system, globalisation is helping to reorganise the structural characteristics of Australian agriculture and rural society. In terms of agriculture, there are strong pressures for farm amalgamations, increased involvement with agri-businesses and other corporate firms, and for greater efficiency and productivity gains. If there is one main finding from the agrifood literature, it is that agriculture in the advanced nations is coupling with corporate capital in ways that are helping to reconstitute both world markets and the farming systems that serve them. Globalisation has also altered the relationship between food and farming. Once, the latter resulted in the former. Now, the former is determining the conditions for on-farm production. This may be viewed positively, with sections of the corporate food industry claiming to be fostering a 'clean and green' and more sustainable agriculture. There is also evidence of this in terms of the marketing of organic products in mainstream supermarkets. In contrast, there is also evidence that farmer involvement with the corporate sector – through purchasing of high-tech inputs, and dependence on finance and production contracts – is leading farmers further down the 'productionist' track, where agricultural activities are having deleterious effects on the environment, thus reducing the natural capital upon which rural Australia depends.

Finally, rural restructuring appears to be depleting more than natural capital. Studies throughout Australia have shown that, while smaller communities show resilience, their future will be based on harnessing the 'scarce resources' of human capital and social capital that are expected to catalyse future development. The steady removal of both is occurring under conditions of the reduction of private and public investment. It has been suggested elsewhere (Dore & Woodhill, 1999; Gray & Lawrence, 2001a) that for regional Australians to be participants in the new global order they must be provided with resources – the social and economic infrastructure required to stimulate an enthusiasm for the future. Planning for sustainability is another important task of government – yet, longer-term planning has been largely dismissed as being interventionist in an era of 'self-help'. The challenge for government, business and community is to establish a suitable framework for local/global interaction in Australia and to address the problems of social fragmentation and social exclusion that have been perpetuated by the neoliberalist drive for agricultural – and wider rural – restructuring.

NOTE

1   This is an extended and updated version of the paper 'Agri-food Restructuring: a Synthesis of Recent Australian Research', published by the Rural Sociological Society in *Rural Sociology* 64(2), 1999, pp. 186–202.

7

THE CHANGING ECONOMICS
OF RURAL COMMUNITIES

Richard Stayner

The purpose of this chapter is to show that economic factors help to explain many of the observed changes in the spatial distribution and condition of rural communities. A diverse range of economic factors, acting on rural places with different histories and endowments of capital, generate different economic outcomes for these communities. The chapter also considers the ways in which these factors might affect the future sustainability of rural communities.

I begin by explaining the ways in which economic changes within the farm sector and its associated industries have altered the distribution of economic activity between rural communities. Changes within the other primary industries, and the effects of changes in other industries represented in rural communities, are then considered. The various categories of capital are examined in turn, to illuminate the ways in which changes in them have affected, and are likely to affect, the economic condition of rural communities. The effects of economic policy interventions on rural communities are then briefly outlined. Finally, conclusions are drawn regarding the role of economic factors in the possible future of rural communities.

At the time of Federation, about two-thirds of Australia's population lived outside the capital cities. Over the next 70 years, rapid population growth in those cities, fed by immigration from other countries and from rural areas, meant that the proportion outside the capital cities gradually fell to just over one-third, since which time it has remained relatively stable. However, as Hugo shows in chapter 4, there has still been population growth within non-metropolitan areas, but with considerable redistribution of population between individual places and regions.

Historically, rural Australia was a place whose role was seen as the exploitation and transformation of natural resources – whether they were farming or grazing lands, forests, fisheries or mineral deposits. The economic and social roles of rural towns reflected that purpose, in that they were essentially service centres for these primary industries. Today, however, the relationship between the economic condition of primary industries and the economic and social condition of rural communities is often much less strong and direct, and factors originating elsewhere in the economy are relatively more important in determining the circumstances of rural communities. Understanding how this economic relationship became attenuated is helpful to understanding the current position of these communities.

ECONOMIC CHANGE IN THE FARM SECTOR AND ASSOCIATED INDUSTRIES

Agriculture contributes to a regional economy through the production of commodities on farms, in 'upstream' processes that take place in the manufacture and supply of farm inputs, and in processes that take place 'downstream' of the farm gate, such as in further processing, marketing and distribution of the transformed commodity. Over many decades, there has been continuous structural change in most of our farming industries as a response to the long-term decline in farmers' 'terms of trade' (the ratio of prices that they receive for their commodities to their costs of production). This structural change has been both land-extensive (larger and fewer farms) and capital-intensive (involving the substitution of capital for labour on those farms). Because employees in the past typically lived and spent much of their income locally, the employment of labour usually generated significant linkages within the local or regional economy. By contrast, the use of purchased capital tends to leave a lighter economic footprint in those communities. Technological change, whether on-farm or in upstream or downstream industries, has been crucial in this transformation.

Over time, there have been significant shifts in the location of value-adding processes, from on-farm or near-farm to regional centres, metropolitan areas or abroad. In the early days of Australian farming, the key resources were land and labour. Many of the non-labour inputs were manufactured either on-farm or locally. Local manufacturing and distribution of both farm inputs and household goods were relatively labour-intensive. As well, scale economies in these processes were not great by today's standards, and local businesses were sheltered from intensive non-local competition both by the costs of the rudimentary technologies of transport, storage and distribution, and by tariffs on imported manufactured goods. Thus, the local economic footprint of the farm sector was significant, indeed dominant.

On the upstream side, for example, horses and men once provided much of the on-farm power. Horses required local labour to care for and work them. Feed sources were usually local, as was the manufacture of farm implements, carriage and wagon building, saddlers, farriers and so on. Compare this with the current sources of power and machinery inputs on farms. Farm machinery is now made in places that are able to reap the considerable scale economies inherent in its design and assembly. Fuels are extracted and processed in relatively few locations that are remote from farming areas. The sales, servicing and repair of farm machinery also exhibit scale economies, as a result of high inventory costs and the increasingly sophisticated skills and equipment required for servicing. Investment in these facilities is only repaid if the customer catchment is large. So, dealers and facilities have become fewer and further between. Likewise, the labour of large teams of men sewing up and carrying wheat bags or cutting sugar cane at harvest time has been replaced by machines. While from the farmer's point of view these machines add more value, much less of the spending that delivers this added value remains in the region. Even where significant amounts of (mainly seasonal) labour are still employed in the farm sector, the local capture of employee spending is reduced by the greater mobility of workers, given advances in transport technology such as cars and roads.

Other inputs display a similar pattern of spatial relationships. Greater use of debt finance by farmers results in the leakage of debt servicing payments from the region, and technological changes in banking have generated economies of size in the provision of banking services that have reduced their presence in smaller centres. As in other industries, knowledge-intensive inputs are increasingly important. Agricultural research, farm chemicals, new plant and animal genetic material, information and communications equipment and services, and other knowledge-intensive inputs all display scale economies in their design and manufacture, and are therefore concentrated in places where such design and manufacture is relatively efficient. These locations are not necessarily, or even mainly, rural. Finally, better communications and transport technologies have made the purchase of inputs from non-local sources more feasible for farmers pursuing the lower costs available from high-volume suppliers. Similar substitutions have occurred in farm households. The labour of farm women and 'hired help' tending the household garden, preserving and preparing food, cleaning, making and washing clothes, and educating children, has been partly replaced by a range of technologies and purchased goods and services.

On the downstream side, there have been analogous technological and other changes leading to scale economies in the further processing of farm commodities, and the concentration of this further processing in fewer, often non-rural, places. For example, technological innovation in livestock selling has reduced the role of many local saleyards, and scale economies in meat processing have reduced the competitiveness of many of the abattoirs in rural regions. Similar factors have been operating in other food and fibre processing sectors. The location and scale of processing facilities are influenced by several factors that confer no special advantages on dispersed regional locations. These include:

- the cost advantages of sourcing supply over a full year
- the need to blend supplies of raw materials from several regions due to seasonality of production and the variability of commodity characteristics
- the relatively low density of production per unit area in broadacre farming industries
- the advantages for a processing plant of locating close to the major concentrations of consumers
- the advantages that larger centres offer in terms of better access to concentrations of high-level skills in product development, design and marketing

In summary, economic factors have tended to result in the concentration of both upstream and downstream value-adding processes in fewer, often non-rural, places.

There are, of course, exceptions to these general trends. Some of the interactions between the farm sector and its associated industries still require close contact and rapid response, and this favours local supply and support. As well, the intensity of local commodity production associated with irrigation sometimes generates a 'clustering' of input suppliers and output processors in a producing region. For example, cotton production in the Namoi Valley of New South Wales now generates an array of relatively knowledge-intensive input suppliers and commodity handlers in Narrabri, while intensive horticulture in the Murrumbidgee Irrigation Area is responsible for continued growth in Griffith.

Meanwhile, the mix of goods, services and amenities that are required by farm households and businesses has changed. Given the growing complexity of operating both farm businesses and households, rural residents need to interact more frequently with an increasingly wide range of businesses, agencies and individuals beyond the farm. Because of the spatial redistribution of economic activities described above, the relationships that farm businesses and households have with other businesses and agencies are now much more spatially diffuse than they previously were. This means that farm households now interact with several communities, perhaps banking and buying farm merchandise and groceries in one, going to the doctor in another, and sending children to school in another. Newly emerging functions are less likely to be found in the smaller towns, since larger towns offer better access to the skills and services that are important in the location decision of new businesses. This further diffuses the traditional relationships that farm people have had with their closest town.

A role that rural communities perform for an increasing number of farm families is the provision of off-farm employment for one or more family members as a way of supplementing farm income. Larger rural towns are more likely to be able to provide opportunities for off-farm employment or business and, since people are likely to do more of their spending in the town where they work, this is another centralising factor that favours businesses in the larger towns.

Does the form of ownership (corporate *versus* family) of farm businesses affect the strength of the economic linkages between the farm sector and the town? Are 'corporate' owned farms more likely to bypass the local town and spend non-locally? In the 19th century, the squatters tended to have their important economic and social linkages with the metropolitan centres, in contrast to the selectors, whose linkages were mainly with the local towns. Today, it is likely that non-local linkages are more important for the larger farm businesses, irrespective of their ownership structure.

ECONOMIC CHANGE IN NON-AGRICULTURAL PRIMARY INDUSTRIES

As with agriculture, the other traditional rural primary industries of mining, forestry and fishing have all seen similar processes of long-term substitution of non-locally sourced capital inputs for locally sourced labour. Goldmining was perhaps the extreme example of this, where originally many thousands of 'diggers' with little in the way of capital soon exhausted the deposits that were easily accessible, resulting in the demise of dozens of briefly thriving mining communities. Only a handful of the once mighty mining communities remain (Mt Isa, Broken Hill, Kalgoorlie), illustrating the way in which communities reliant only on the exploitation of natural capital (an ore body) are unsustainable. In more recent times, a number of 'company towns' were created to mine iron ore in the Pilbara region, but these did not usually develop a wide range of community functions or a diversity of economic opportunity.

A range of factors – the gradual substitution of capital for labour, increasing volatility of resource commodity prices, a range of problems relating to the creation of 'healthy' single-purpose communities in very remote areas, more cost-effective air transport, and changing preferences of miners and their families – have combined to make it more attractive for mining companies to supply labour to mine sites on a 'long-distance commuting' or 'fly-in/fly-out' basis. This is an extreme form of the separation of place-of-work from both place-of-residence and place-of-spending that has occurred in many rural communities.

Forestry-dependent communities have experienced similar changes resulting from both the exhaustion of the natural resource and the advent of more efficient and capital-intensive methods of silviculture and timber processing. This has resulted in the closure of smaller mills and the concentration of processing capacity in fewer and larger mills. Conflicts over the desirable ways of using and replacing the natural resource have also been important in determining the spatial redistribution of forestry and its associated economic activities in rural communities. As in mining, low-skilled jobs are being replaced by relatively high-skilled ones, and this has changed the structure of employment opportunities in many rural places. Similar forces (capital-labour substitution and the concentration of processing capacity) have been at work in the fishing industry. The pressure that has been put on the sustainability of the natural resource has attracted policy intervention to close some commercial fisheries to all but recreational fishers. The future effects on communities where fishing is important could be significant.

ECONOMIC CHANGE IN OTHER SECTORS

While changes within the traditional rural sectors and their upstream and downstream industries have been important, the economy as a whole has also been restructuring, in general away from employment in primary and secondary industries and towards employment in tertiary (services) industries. While there has been some growth in the latter in rural areas, they have not in general shared in the growth in the services industries to the same extent as metropolitan areas. These industry sectors tend to be knowledge-intensive, and rural areas have systematic shortfalls in levels of skills and in the renewal of skills. Rural areas consequently have low representations of the high-growth industries.

A modern knowledge-intensive economy displays 'agglomeration' effects – the tendency for similar or associated businesses to cluster spatially – based on the importance of access to rich and diverse pools of skills, business services, networks of associates, information sources, ideas and creative opportunities. This gives them access to some of the advantages of scale that are not available in smaller rural places (Krugman, 1991). Similar forces, especially technological changes leading to scale economies, and improved access to non-local providers, have been operating in other industry sectors represented in rural regions, such as retailing, medical services, education and recreation. Consequently, these high-growth sectors tend to concentrate in the larger rural towns.

The restructuring of the national economy has been accompanied by net growth in jobs, in the sense that job creation in growing industries has more than offset job losses in others. Regional economies have been restructuring too, except that in many regions, even where output has increased, job growth has been less than enough to offset job losses, because of the types of jobs gained and lost. As well, the costs of restructuring, in terms of disruption to the lives of the people involved, are higher in rural than in metropolitan areas, since relatively undiversified local economies offer fewer alternative job opportunities in the growing industries, and the spatial separation of job markets means that people have to travel further, or move their place of residence, in order to find another job. As well, the lower levels of education and skills in rural areas make it more difficult for those who lose their job to acquire new skills.

Tourism and road transport services are two growth sectors of the economy in which rural regions have a reasonable representation. On the other hand, the public sector service industries that have traditionally been relatively important employers in rural areas – government, education and health – and the recently corporatised or privatised electricity and telecommunications utilities, have been experiencing their own adjustment pressures, some of them policy-induced.

CHANGES IN CAPITAL IN RURAL COMMUNITIES

I now review briefly some of the major changes that have been occurring in the several forms of capital that contribute to the economic and social condition of rural communities. The use of a 'capitals' framework is also discussed in chapter 1 and in Cocklin and Alston (2003).

HUMAN CAPITAL

Human capital is the personal capacities of *individuals*, including their knowledge, skills and general abilities (including their health). While economic analysis once dealt with people merely as 'labour', it now recognises the qualitative characteristics of labour as 'human capital', and acknowledges its crucial influence on economic outcomes.

Most inland rural areas have been experiencing gradual reductions in population over the last 25 years, with the smaller places most heavily affected, but with some regional centres and nearby places experiencing growth. Out-migration of young adults and families with children has particular implications for the local economies of rural communities. Rural communities are, in general, under-represented in the early adult age groups and over-represented in the elderly age groups. Since skilled people tend to be more occupationally mobile, there is a tendency for those with higher levels of human capital to leave the area when restructuring occurs, taking with them skills and networks that are important to the networks of social capital as well. On the other hand, some rural communities have been attracting low-income and single-parent families drawn to cheap housing. This trend may now be spreading to relatively well-off people in the older age groups. Some rural communities, especially those within relatively close range of metropolitan centres, are now reporting strong demand for housing from metropolitan areas. This effect no longer seems restricted to coastal areas. (See chapter 4 for a fuller discussion of these demographic trends.)

Given the increasingly skill-intensive economy, it is important that rural places continue to be able to attract those with economically valuable skills and knowledge. In some rural communities, with an attractive mix of social, cultural and environmental amenity, there have been enough of these in-migrants to make a significant impact on the local economy, especially if they start up or buy businesses with some prospect of growth. There may be little that most rural communities can do to stem the outflow of young people who leave in order to seek educational and early career opportunities. It will be important, however, for these communities to regain at least some of those who leave for those reasons, and there are signs that this is happening. Nowadays, however, many skilled people, such as doctors, have partners who also require satisfying work. It is difficult for the smallest rural communities to offer such employment opportunities.

The renewal of human capital in the farm sector (that is, the recruitment of new farmers who have appropriate skills and motivations) is inhibited by: social and cultural factors relating to the ways in which farming has traditionally been an inherited occupation; by values that farm families hold regarding the importance of retaining ownership of their land; and in general by the problems of negotiating important stages in the farm family life cycle (entry, managerial succession and retirement). These values may be weakening, however, and they are not essential to the continued economic performance of the farm sector. Local schools and other institutions for building and renewing human capital will remain crucial, by providing retraining opportunities, and by developing a culture of higher educational aspirations and continuous learning in rural communities.

PRODUCED (OR BUILT) CAPITAL

Produced or *built* capital refers to the 'built environment' and anything else that has been made by humans, whether in private or public ownership. This includes the physical assets of businesses and households, as well as public physical infrastructure. The three major forms of produced capital that affect the economic condition of rural communities are farm business capital, non-farm business capital and public or community capital.

It has been argued above that the farm sector is unlikely to be the driver of sustained future economic growth in most rural communities, since in most regions the economic linkages between the farm sector and rural places are becoming more spatially diffuse and attenuated. At the same time, the average rate of return on farm business capital is typically below that earned by other businesses, and its variability over time is higher. What, then, are the prospects for transferring some of the capital tied up in the farm sector to other economic ends in rural communities? A high proportion of farm business capital is immobile, since it is tied up in land and, as noted above, farm families typically prefer to pass this asset on to the next generation. As well, many farmers (and other rural residents) with funds to invest seek to diversify their assets by investing outside the local community.

While there are no reliable data on flows of business capital, in view of the previous discussion of the processes that are redistributing economic activity it is plausible to suggest that private non-farm capital is tending to flow away from the smaller rural towns towards the larger regional centres, coastal and metropolitan areas, in line with the relative shift in economic activity and opportunity. The agglomeration economies enjoyed by businesses in metropolitan areas and the larger regional centres give those places a competitive edge in attracting capital, and there is also a lack of attractive business opportunities in many rural communities. This is a counter-argument to the occasional proposal that superannuation funds be required to invest a proportion of the funds contributed by their rural members into projects in rural regions. Since the deregulation of the finance markets in the mid-1980s, it is more difficult to argue that 'market failure' inhibits the flow of capital into worthwhile projects in rural regions. It is possible, though, that the relative lack of large projects with the right mix of risk and return means that major institutional investors do not pay much attention to investment opportunities in rural regions.

Public investments in produced capital – the provision of public infrastructure, such as railways, irrigation and electricity generation schemes – were once used as the catalyst for attempts to generate economic development in rural regions (see Davison, in chapter 3). But governments these days tend not to subscribe to the belief that 'if we build it, they will come'. Stricter financial disciplines are now imposed on many government business enterprises, and public infrastructure investments are subject to much closer scrutiny and criteria that more closely resemble those of the private sector. While public interest and equity criteria may also be considered, in general the trend favours the larger centres over the smaller ones.

NATURAL CAPITAL

Natural capital refers to:

> ... the renewable and non-renewable [natural] resources which enter the production process and satisfy consumption needs, as well as environmental assets that have amenity and productive use, and are essential for the life support system. (ABS, 2004)

Most of Australia's rural communities were established to serve industries that extracted, exploited or in some way modified the stocks of natural capital. In the case of mining, this extraction sometimes led to the rapid demise of a community as the ore body was exhausted, unless progressively more efficient methods for extracting the ore extended its life.

In the case of agriculture, the often radical modification of natural systems led to the development of farming and the communities that depended upon it, at the cost of eroding the stocks of natural capital in the form of reduced biodiversity, species extinction, and decline in soil and water quality. Occasionally these modifications led to the rapid collapse of farming systems and of the social systems that depended upon them (such as post-First World War soldier settlement in the Mallee). The effects of land uses that have overstressed the natural systems are now becoming more obvious, such as dryland soil salinity, particularly in the Western Australian wheat belt, and irrigation-induced salinity in the southern Murray-Darling Basin. The effects of declining soil and water quality are being felt not only within farming systems but also in rural towns; for example, in the form of damage to produced capital such as roads and buildings, and in effects on town water supplies (Tonts & Black, 2003).

While vast areas of the nation's natural capital are under the primary control of farm businesses, albeit in a highly modified state, there remain considerable natural resources under public ownership. The potential for these to be the basis for economic activity, through tourism and other non-consumptive uses, is increasingly being recognised. Indeed, private landholders are also recognising opportunities for the development of such enterprises.

In general, the relationships between the changing condition of these stocks of natural capital and the economic condition of the rural communities that depend upon them are not well understood, and are sometimes indirect and long delayed. This makes it difficult for communities to respond to the problem, since the objective of sustaining the short-term economic and social condition of the community conflicts with the (usually) longer-term objective of retaining the capacity of the natural systems to deliver both consumptive and non-consumptive use values indefinitely.

SOCIAL CAPITAL

Social capital refers to the networks, shared norms, values and understanding that facilitate co-operation within and between groups (ABS, 2004). As with natural capital, the measurement of stocks of social capital presents practical difficulties (ABS, 2004), but its economic significance is now accepted (Productivity Commission, 2003a). As discussed in chapter 1, at least two important forms of social capital are now recognised: *bonding* capital refers to the trust and social cohesion among people who interact relatively frequently, often because they share a place and its common local concerns and norms; while *bridging* capital refers to the co-operation that results from networks of relatively dissimilar people in different places, occupations or levels of power.

Conventional images of small rural communities implicitly assume that they have relatively rich stocks of bonding capital, arising from their social familiarity, cultural similarity and sense of shared fates. This is becoming less true for many small communities, as a result of the weakening of local linkages as residents now interact with more geographically diffuse businesses, agencies and individuals. These increasingly diffuse relationships between farm families and rural communities may be weakening the traditional social linkages that these families have had with 'their' town. If so, it could be weakening the quality of rural social capital (through a reduced willingness and time to be involved in voluntary or collaborative activity), and therefore further eroding the capacity to produce non-market goods and services in some rural communities. As well, collaborative community action, from running a sporting club or bush fire brigade to preparing a funding submission, now requires more formal skills and procedures, and those individuals with the formal and personal skills to conduct these tasks are becoming very thinly stretched across several groups in many communities. Furthermore, those who are the important nodes in the social capital network are often occupationally mobile, and may only be in the one place for a short time.

The importance of social bridging capital to the adaptability of rural communities is increasingly apparent. To the extent that communities are now acting more like nodes in regional networks of communities that together meet a much broader range of needs than any one community can, the strength and diversity of *non-local* networks, sense of shared *regional* fates and trust between non-intimates become much more important. A valuable source of bridging capital is an inflow of people who already have contacts and familiarity with individuals, businesses and agencies elsewhere. This underlines the importance for a community of being able to attract at least some of these well-connected individuals. Some communities, especially those with transport access to a metropolitan area, with relatively attractive environmental and social amenities and with adequate educational, medical and recreational services, are attracting more of these people than the more remote communities.

The several kinds of capital, when combined, generate a wide range of social outputs that are important to, and valued by, a community. As economic and social life becomes more sophisticated and complex, there is a corresponding increase in the range and complexity of the social outputs that people need or aspire to in order to manage their lives. In general, small rural places have a declining capacity to provide a full range of this expanding set of functions. Thus, as well as losing some of their existing functions, they typically do not acquire all the newly emerging ones. In other words, they can no longer provide (if they ever did) for all the needs of their residents.

INFLUENCE OF ECONOMIC POLICY

Historically, the spatial distribution of rural communities has been influenced by deliberate policy as much as by market forces. Land settlement policy in the 19th century was a response to 'land hunger' following the gold rushes, especially the need to absorb workers displaced from the diggings, as well as a response to the *ad hoc* settlement of vast areas (squatting). As Davison shows in chapter 3, the policy envisioned a matrix of 'closer settlement' of farms, dotted with rural communities that would be sustained by their relationships with agricultural production. Because of the small size of the domestic market, production was soon oriented towards export. Hence, producers of these commodities soon learned about the inherent variability of their economic wellbeing, as commodity price booms and crashes occurred almost from the start of settlement.

Echoes of these closer settlement policies are evident in the Soldier Settlement Schemes that followed both World Wars, and in irrigation developments throughout the 20th century. Following both World Wars, temporarily buoyant commodity prices led authorities to underestimate the area of land and the amount of financial capital that farmers would need to generate economically sustainable farm businesses over the long term. Thus, policy tended to overestimate the capacity of the land to support intensive agricultural production, and hence the capacity of rural regions to support and sustain rural communities.

The period immediately following the Second World War was another of considerable agricultural expansion, aided by closer settlement schemes, irrigation development and government support for co-operative marketing schemes for many commodities. Meanwhile, a policy of high import protection for secondary industry put heavy imposts on the export-dependent primary sector, even though some rural communities acquired highly protected manufacturing industries, such as textiles, clothing and footwear. Measures aimed at decentralising economic activity to rural regions were tried, but with limited success.

Australia's primary industries have always been exposed to international influences. In the first two-thirds of the 20th century, policy aimed to shield producers from these influences by means of commodity price support and fixed exchange rates. As the economy gradually developed other strengths, and agriculture became a smaller part of the economy, policy needed to shift away from a protectionist stance. This involved a combination of policies aimed at exposing the economy even more to international influences, such as floating the exchange rate, deregulating international capital flows and reducing industry protection. (See, for example, Herbert-Cheshire & Lawrence, 2003, and Dibden & Cocklin, 2003, for commentaries on dairy deregulation.) These resulted in increased flows of debt capital into the farm sector, as the banks, newly exposed to foreign competitors, scrambled to lend to farmers. When commodity prices experienced one of their periodic crashes in the mid-1980s, and the deregulated interest rates rose to unprecedented levels, severe financial stress was experienced by many farmers, especially those who took loans in foreign currencies, whose risks they had no experience in managing. Both farmers and their lenders have taken some time to learn how to manage the greater volatility of interest and exchange rates. The subsequent closures of bank branches in many of the smaller rural communities probably had little to do with any losses suffered by the banks on their rural lending at that time, but most farmers are now much less loyal to their traditional local banks than they once were. Dibden and Cocklin (2003: 184) show the damaging consequences of bank rationalisation on social capital, particularly relationships of trust.

At the same time, a wide range of microeconomic reforms was targeted at various sectors of the domestic economy, aimed at improving their international competitiveness, but these imposed significant adjustment costs on certain businesses and places. There was a perception that rural regions fared relatively poorly in the distribution of benefits and costs of these policies. The Productivity Commission (1999: 283) found that the incidence of benefits and costs across country areas was more varied than in metropolitan regions, but that rural regions as a whole stood to benefit from national competition policy. They added that the effects of these reforms on most, but not all, regions were likely to be less significant than those resulting from the broad economic forces that continually reshaped economic and social conditions in rural regions.

The policy settings of the Commonwealth government towards rural Australia have mainly been found in policies for rural *industries* (dominated by economic efficiency objectives), underpinned by generic *social* (welfare) policies targeted at individuals and families regardless of where they live. Social security payments and Commonwealth revenue sharing grants to local government via the Grants Commission are the most important manifestations of the latter. In the past decade or so, there have been relatively few programs that pay specific attention to rural *places* as such, although this has not always been the case. There has recently, however, been an increasing number of experiments by Commonwealth and state governments in place-specific programs (see chapters 11 and 12).

Governments have started to respond to the strong expressions of dissatisfaction of rural people concerning the decline of social and economic fabric in rural communities by attempting to develop programs for the reversal, or at least the amelioration, of the observed trends. In these responses, governments are increasingly calling on the presumed self-help capacities of rural communities for them to become 'partners' with government in developing and delivering programs for rural communities. In many rural places, however, the depth and breadth of resources, particularly the human and social capital, may not be sufficient to allow communities to respond to, and participate effectively in, such programs.

The policy choice can perhaps be paraphrased as 'jobs to people *versus* people to jobs' or, as Bolton (1992) put it, 'place prosperity *versus* people prosperity'. In other words, a distinction is drawn between, on the one hand, policies that focus on enhancing the capacity of places to meet the needs of local people and, on the other hand, policies that respond to the needs of individuals, regardless of their location. The latter include policies for enhancing the capacity of the national economy to generate jobs, albeit in places that might require people to move to them. There is no doubt that the latter focus dominated rural policy in the period of dramatic reform of the Australian economy from the mid-1980s to the late 1990s. But a policy environment that places almost exclusive emphasis on the mobility of people required for the efficient operation of markets risks overlooking, and therefore diminishing, important place-dependent values. Should such values be considered in the development of policies that address the problems of particular rural places?

Bolton (1992) invoked 'sense of place' as one factor that could justify the use of place-specific policies for rural development. Noting that social capital is productive and has local public goods aspects, he observed that the 'critical question for state and national policy is whether the "publicness" extends over a wider range of space than the community itself' (Bolton, 1992: 193). This raises three questions, according to – and adapted from – Bolton (1992: 193):

- How valuable is the social capital in particular places for the larger region, and for the nation?
- Should, and do, people in Sydney care whether a strong sense of place exists in other towns in New South Wales, or in a Pilbara mining town?
- Which *particular* places (of the many hundreds that are in similar circumstances) should attract the scarce public resources to sustain their communities?

If the sense of place is a valuable social asset for the larger region and nation, what are the appropriate roles for state and national governments? Is the value sufficiently high to justify government action? Are there appropriate policy instruments?

Does reliance on 'people-prosperity' policies, to the exclusion of place policies, allow the social capital to erode at a rate that is too rapid for the region or nation?

CONCLUSION

There are over 1600 towns of between 200 and 50000 people in Australia, some of them within the metropolitan fringe, and many of them in coastal regions. Australia has only eight medium-sized cities of between 50000 and 100000 people, and one of these is a Territory capital (Darwin). Of the cities over 100000, only Canberra is inland. Given the economic factors making for increasing returns to scale for communities (that is, a place is more likely to grow if it is big already), inland Australia would appear to have few rural centres with a high inherent potential for population growth. The key brute economic facts of inland rural Australia are its relatively low densities of population and production, and its disadvantage compared with the capacity of the cities to offer access to dense networks of knowledge-rich resources that are increasingly important to new economic activity.

A recent topic in economic development theory has attracted considerable attention for the insights that it may offer for Australia's rural regions. Porter (1998) stressed factors that stimulate productivity growth in the economy at large, and emphasised the role of industry *clusters*, which are critical masses of firms and supporting institutions in a particular location. In Porter's analysis, the prosperity of clusters depends less on traditional scale economies and more on synergies with other nearby firms and, significantly, on investment in human capital and networks. Porter's framework suggests that governments can stimulate regional economic growth by facilitating the development of clusters through ensuring that human capital resources, information and transport infrastructure, and other productivity-stimulating inputs are present.

It is difficult, however, to identify more than a few inland rural regions in Australia that have a critical mass of related firms and supporting infrastructure comparable to the overseas examples of industry clusters that inform Porter's model. The closest approximations may be softwood production and processing in south-eastern South Australia and south-western Victoria, horticulture in the Murrumbidgee Irrigation Area, and cotton in the Namoi Valley. It is notable that two of these are irrigation-dependent regions. It is in these that conflict is likely to continue between the growth of these industries and the sustainability of the services of natural capital for both the farm sector and for non-consumptive uses.

The social and economic conditions of rural communities will continue to be influenced, as they always have been, by national and international forces beyond the control of those communities. Closer to home, economic outcomes will also depend not only upon the stocks of the several forms of capital over which communities do exert some control, but on how these resources are combined. Capitals act as complements rather than substitutes, and all are important (Cocklin & Alston, 2003). Nevertheless, it is clear that the attraction and retention of *human* capital and the nurturing and enhancement of *social* capital are increasingly crucial to the future of rural communities (see Smailes & Hugo, 2003). The creation, attraction and retention of these in turn depend on the maintenance of environmental and cultural amenity. Without these, no amount of built capital is likely to result in sustained economic growth in rural communities. As well, 'human agency' is important. That is, it is important that community leaders make good decisions regarding how the limited resources of their communities are used, and what aspects of their community's needs are given priority. The challenges for rural communities are:

- how to choose which functions will be provided locally
- how to maintain, create or import the various forms of capital that combine to meet community needs
- how to get the most out of the limited resources that they do command

The history of the failure of primary industries to create economically and socially self-sustaining communities in rural Australia suggests that these will not be the engine of future growth for rural communities. Fortunately, rural regions are now the setting for the expression of a much wider range of aspirations and values than was contained within the historical vision of 'nation building' via the extraction or modification of natural capital. Examples of these values include:

- the value of non-metropolitan Australia as the site for the evolution and expression of a stewardship ethic towards our natural resource endowment
- important symbolic values and identities
- its value in creative and spiritual generation and regeneration, such as in the arts, literature and film, and in the direct experience of nature and space
- values attaching to native flora and fauna
- various tourism values
- scientific, educational and research values; for example, archaeological sites and remnants of flora and fauna of world significance

Some of these values offer opportunities for economic development, while others do not. There are disagreements, sometimes sharp, over the weight that these alternative values should be given. The challenge is to design economic and other institutions that reflect, balance and protect these values. In the context of this chapter, the question that arises is: What will be the roles, and the economic condition, of rural *communities* within this evolving vision?

8

GENDER PERSPECTIVES IN AUSTRALIAN RURAL COMMUNITY LIFE

Margaret Alston

Writing about rural communities in 1994, Whatmore, Marsden and Lowe (1994: 1) noted that:

> The danger of the marginalisation of gender issues is undiminished, at least in the advanced industrial countries ... Here the significance of gender relations in the social and economic dynamics of rural life continues to be segregated as a specialism in which rural women and 'their' issues can be safely corralled by researchers and policy-makers, while mainstream research and policy concepts and concerns proceed untouched.

This wisdom still holds true in Australian rural studies, with gender relations still largely an add-on issue for many researchers and policy-makers or confined to feminist researchers and rural women's policy units working on women's issues. Yet, when conducting research in rural communities, it is evident that gender is a defining feature of rural community life (Alston, 1995; Alston, 2000; Wilkinson et al., 2003; Alston & Kent, 2001; Alston et al., 2001a). One has only to examine the profile of the rural landholders and business owners, the holders of public office, the professional class, the users of public space, the decision-makers, the news-makers, the voluntary workers and the child-carers to note that gender is a critical determinant of the lived experience of small-town rural Australians. The division of labour, the uneven power relations and the various ways in which men and women relate to their rural environment confirm that there are processes at work that make the study of gender critical to an understanding of rural community life.

Gender is not a category that is biologically determined. Rather, it refers to the socially constructed societal expectations of women and men. Yet, the slipperiness of the concept is evident. It is not static or ahistorical, but responsive to changes wrought, for example, through rural restructuring and globalisation. Nonetheless, it is arguable that responses to these events remain gendered (Gray & Lawrence, 2001a). Thus, while we witness women's greater access to the labour market and the crisis in agriculture as significant recent structural changes that shape and reshape rural gender relations, we note that new but equally gendered alignments result. It is therefore imperative that gender, its intersection with rurality and the discursive practices on which both are reliant be understood and incorporated into our analysis of rural communities.

In this chapter, I examine the concept of gender in an Australian rural context, using the framework of the five capitals – economic, institutional, human, social and environmental – to show that living in small Australian rural communities in the early 21st century is very much a gendered experience.

GENDER ORDER/GENDER REGIME

In explaining the resilience of gendered arrangements, Connell (1987, 2002) notes a distinction between what he terms the *gender order* and the *gender regime*. The gender order he describes as the wider patterns in society that exist over time or the macro-level historically structured power relations between men and women. On the other hand, he views the gender regime as the patterns in smaller units, such as the family, community or institution, that may mirror the gender order or may alter over time, creating the possibility of change in a wider sphere. The gender regime is the micro-level gender politics experienced by individual women and men in their homes, workplaces and communities. McNay (2000) makes a further distinction between the micro level (at the individual face-to-face level of interaction), the meso level (within organisations, communities and ethnic groups) and the macro level (the wider societal level, such as political and economic systems). What both agree on is that if we are to understand gender and the unevenness of gender restructuring, we must look at the interactions at different societal levels.

The gender relations of rural community residents are shaped by the macro-level gender order and the discursive practices that affect an individual's response. Thus, for rural women and men, the negotiations that they undertake at the gender regime level of the family, farm and rural community are shaped within the constraints and expectations of the gender order of women's subordination (O'Hara, 1994). Hence, while negotiations within the gender regime can suggest co-operation and self-sacrifice, they can also contain elements of exploitation, coercion and struggle (Davidson, 1991; Beechey, 1987; Whatmore, 1991). Yet, as Connell (2002) notes, gender would not exist if gender relations were not actively reconstituted in everyday life, facilitated by institutions such as the media, the church and the market. Thus, attention to gender relations is ongoing and active, and a thorough understanding of gender demands acknowledgment of the processes through which gender relations are continually reconstituted.

In coming to grips with gender and changes in gender relations, Connell (2002: 142) also refers to the *patriarchal dividend* to explain the investment that men as a group have in maintaining an unequal gender order; he argues that this dividend results in monetary benefits as well as greater 'authority, respect, service, safety, housing, access to institutional power, and control over one's own life'. Thus he argues that the patriarchal dividend, or what Oakley (2002: 221) irreverently refers to as the 'penis privilege', is worth defending and so the rationale for the maintenance of the gender order is evident.

HEGEMONIC MASCULINITY

In rural areas, the maintenance of the gender order results in a dominant masculinity 'that occupies the hegemonic position in a given pattern of gender relations' (Connell, 1995: 74). Hegemonic masculinity (Connell, 1995, 2002) allows not only male power and influence to be superordinate, but also one view of the world to be normalised and projected onto the subordinated, in this case women, as natural. Hegemonic masculinity is evident in rural Australian communities when we look at the institutional structures, such as the law and religion, and the processes and practices, such as patrilineal inheritance, through which gender relations are constructed and reconstructed in everyday life. Yet, as Laoire (2001) notes, the dominant position of rural men is not unproblematic. For some rural men, their occupation of positions of authority is precluded by their poor educational qualifications and by their low-paid, casual or temporary employment. On the other hand, masculine control of resources and land in communities reliant on agriculture is facilitated by, and facilitates, hegemonic masculinity, giving greater influence to some men.

In many small inland rural communities, agriculture is still the dominant industry. The control of the resources of agriculture ensures that men have greater influence in industry and in communities reliant on agriculture. The practice of patrilineal inheritance has made male ownership and control of land and the resources of agriculture the norm, giving male landholders enhanced prestige and influence in small communities. Dempsey (1992) suggested that less than 5 per cent of farms were inherited by daughters. The same applies to town businesses that are passed on through patrilineal lines. As a result, rural power and influence are passed on predominantly to men. Women come to agriculture and other family businesses largely through marriage, ensuring that their position in the gender regime of farm and rural community life, particularly when there are intergenerational arrangements, is marginal. The low level of female inheritance of farm land and the domination of local government in small inland rural areas by middle-aged male farmers (Gray, 1991) ensures that the public face of power in rural communities remains male. Women make up less than 30 per cent of elected local government representatives, 15 per cent of mayors and 10 per cent of senior local government executives (ALGWA, 2001). As a consequence, a masculinist understanding of rural community life is privileged.

As Foucault (1980) notes, with power comes knowledge and the ability to create or construct a certain discourse of 'truth'. Industry and community knowledge is constructed at sites of power and certain 'truths' become privileged (Campbell, 2000). In rural areas, such sites for the legitimation of hegemonic masculinity include farm organisations and local government, the pubs (Campbell, 2000) and areas such as livestock saleyards. Other sites of masculine power and privilege include public space in rural communities. Often, large areas are taken up by playing fields for male-dominated sports (Alston, 1996). In research on gender in rural sports reporting conducted in 1996 (Alston, 1996: 38), a small-town local government official noted:

> You'd have to say that the money we spend on maintaining and providing sporting facilities would be used by a greater percentage of men than women and the only reason that it is, is because ... the main winter sport is football and the main summer sport is cricket.

Through such rationalisations, the greater resourcing of men and their use of public space is endorsed.

As Campbell (2000) notes, these male-dominated sites – farmer organisations, local government, pubs, saleyards and sporting fields – are significant sites of power where knowledge or certain 'truths' are developed and transmitted, ensuring that a male view of the world dominates. Female-dominated public space in rural communities is limited. During much of the 20th century, the Country Women's Association (CWA) created women's spaces in country towns through their funding of Baby Health Centres and Rest Rooms. In the 21st century, there are few such areas that could be termed public women's space where women can create alternative discourses. Although neighbourhood centres could fall into this category, they lack the resources and widespread attraction of male-dominated sites. There are also gatherings of women for craft activities where alternative discourses are created. Again, however, they lack the public visibility and legitimacy of male sites.

The tacit support of women in rural communities for hegemonic masculinity can be understood through reference to Bourdieu's notion of *habitus* (Bourdieu, 2001). This is defined as 'the historical and cultural production of individual practices … and the individual production of practices' (Schirato & Danaher, 2002), or the way in which individuals absorb and reproduce certain values and customs implicit in their cultural history. Hence, women and men respond to cultural cues, absorbing positions and values that shape their responses to the gender order, their negotiations at the gender regime level and, ultimately, their gender identities. Nonetheless, in recent years, constant scrutiny of male hegemony, and women's individual responses to it, have led to a perceived crisis in traditional masculinity. A new awareness about gender relations and discursive practices has led to what Connell (1995: 226) describes as the erosion of the 'legitimation of patriarchy'. Laoire (2001) argues that what has come to be termed the rural crisis – the loss of faith in agriculture, the changing population dynamics and reduction in services and support – is related to a crisis of hegemonic masculinity. She argues that, despite the fact that men still dominate in rural society, there are fewer sites for patriarchal dominance and, at the same time, women have increased options for resistance. While there is a certain truth in this, for those committed to remaining in rural areas, hegemonic masculinity still dominates and constrains gender negotiations.

Yet it is important to note that hegemonic masculinity, with its implications for subordinated women, is constantly open to challenge and is a site for struggle. Women and men respond in various ways. Many young women leave rural communities for educational and employment opportunities, arguing that hegemonic masculinity, or what they describe more aptly as 'the macho culture', is too great a constraint (Alston & Kent, 2001). Similarly, many young rural men are turning their backs on a life in agriculture, preferring to escape the rigid prescriptions of rural masculinity. Through such resistances is change facilitated. Yet, dominant rural discourses continue to ensure that cultural practices and institutional structures in rural areas combine to privilege men, ensuring that traditional masculine identities are highly resistant to change.

MAINTAINING THE GENDER ORDER THROUGH DISCURSIVE PRACTICES

Foucault (1980) notes that knowledge and power are facilitated by discourse. In rural communities, discourse is shaped around hegemonic masculinity. Using a Foucauldian analysis to understand how meaning is constructed in agriculture, Liepins (2000c) notes that certain positions and knowledge bases privilege men and constrain women, and are evidence of the contestation of power relations. The adoption of these discourses by government and industry leads to a partial and gendered understanding of agriculture that marginalises women.

While discursive practices normalise patriarchal relations, institutional patriarchal structures reinforce them, and these structures are highly resistant to change. For example, a lack of childcare facilities in rural communities and the expectation that women will undertake this role, together with a decline in access to services for the aged, the disabled and the mentally ill, have left a greater burden on rural families in general and rural women in particular. There is limited public discourse in rural communities on carework and the gendering of this role, despite the fact that there are few childcare and disability support services in small rural communities and despite the looming crisis of care brought about by the ageing of the population (Alston, n.d.). Additional evidence of discursive practices that support hegemonic masculinity is not hard to find. Liepins (2000c) notes that the media assists in the construction of farmers as 'tough', 'strong' and 'active', legitimising men's activities and marginalising women, despite their contribution to farm income. The way in which women are represented in rural newspapers' advertisements and text

(Liepins, 2000c; Bell & Pandey, 1989; Macklin, 1995) and the almost total absence of women from rural media reports of sport (Alston, 1996), all indicate discursive practices that shape gender representations and endorse masculine hegemony.

Illustrations of the pervasiveness of a gender order that subordinates women, the dominance of masculine hegemony, the institutional practices and discourses that reinforce these arrangements, and the resistances of rural women and men, are provided in recent research on Australian rural communities. In the rural places studied for the 'Community Sustainability' project (Cocklin & Alston, 2003), women and men were seen to negotiate gender relations against a backdrop of gender order arrangements that privilege some and subordinate others. The way in which gender issues are addressed in these rural communities is discussed below, using the 'five capitals' framework.

ECONOMIC CAPITAL

Agriculture remains a dominant industry in inland rural Australia. Much of the wealth of these communities is tied up in the economic resources of agriculture and much of this is controlled by men. In their study of the Gilbert Valley in South Australia, Smailes and Hugo (2003) note that the mean period of residence for rural males is 22.7 years and for females is 17.4 years. This is explained by the shortage of locally available female marriage partners and the need for in-migration of women. Smailes and Hugo (2003) also note that about a quarter of males, but only 5 per cent of females, have spent their whole lives in their community, suggesting that it is women who are more socially mobile, leaving small communities for higher education and better employment prospects (Alston, 2002). The gendered experiences of rural community residents are already evident in the relative mobility of women and men.

While there are many intergenerational business partnerships, the position of daughter-in-law (the in-marrying partner) remains one of the most precarious in the farm family and family small business in terms of resource ownership and decision-making capability (Alston, 1995). Nevertheless, rural restructuring and the entry of women into the labour market have opened up possibilities for women to renegotiate their position and work role in the family arrangement. Many now work on farms, replacing economically unviable hired labour. For example, only 13 per cent of broadacre and dairy farms hire permanent or casual employees, with 62 per cent of labour contributed by the farmer and spouse (Garnaut & Lim-Applegate, 1998), and farm women and men spend approximately the same amount of time working – 60 hours per week (Garnaut et al., 1999).

Despite resource ownership and control being highly gendered, many farms are no longer sustainable without off-farm income. In fact, in Australia generally, at least 50 per cent of farms are reliant on outside sources of income (Alston, 1995) and in 80 per cent of cases this is provided by women working in nearby rural communities (Elix & Lambert, 1998; Society of St Vincent de Paul, 1998), making the work of women one of the most important aspects of family farm survival. The off-farm work of Australian farm women doubled between the 1980s and the late 1990s (Garnaut et al., 1999) when off-farm work realised 38 per cent of the farm family income, rising to a peak of 83 per cent during 1991–92 (ABARE, 1999) and settling to 63 per cent in 1994–95 (Garnaut & Lim-Applegate, 1998). What appears to be happening is that, while economic capital, at least in agriculture, is still largely controlled by men, women provide an enabling role, allowing men to remain in farming and facilitating the practice of patrilineal inheritance, a subordinate position not restricted to Australian farm women (see, for example, Shortall, 2002; Eikeland, 1999).

Nevertheless, in recent research in small rural communities, change and resistances are evident. New possibilities are opening up that are allowing women to move beyond a subordinate role in economic production. Tourism is one area in which women are becoming small business owners in their own right, challenging established orthodoxies and creating a more prominent role. Additionally, 'new farmers' (Wilkinson et al., 2003), with equitable business relationships between male and female partners, are moving into areas where traditional farming practices have been dominant, establishing new industries such as wine-making. The move to convert Monto, in Queensland, into a cyber-town (Herbert-Cheshire & Lawrence, 2003) is another example whereby new industries may provide opportunities for destabilising masculine hegemony.

INSTITUTIONAL CAPITAL

Many Australian inland towns have experienced a significant and deleterious loss of institutional capital. Latter-day neoliberal policies in support of smaller government, and the pre-eminence of market forces, have led to a withdrawal by governments and non-government organisations from service delivery. Policies of rationalisation, regionalisation and centralisation have led to services relocating from rural communities into regional centres or capital cities. Meanwhile, policies of privatisation and tendering out of services have had a significant impact on rural communities where some services have been reduced under this arrangement and others have multiple and competing providers. Services such as banks, post offices, court houses, education facilities, welfare and medical providers have all been lost to varying degrees in many small rural communities.

While all residents are impacted by these losses, there is a notable differential impact on women and men in small rural communities. A significant example of this is the loss of the obstetrics services in many small communities because of a lack of doctors and insurance liability issues; Stayner (2003) found that no babies can be born nowadays in the Guyra Hospital, meaning that women must structure their lives and families around the need to exit the town during a particularly stressful period. The threat of early labour ensures that many women will leave these communities well before their baby is due, allowing them to avoid a medical emergency, but significantly increasing the disruption and the financial commitment involved in having a baby. For those unable to afford a timely departure, significant stress arises around the time of birth.

Giving birth is but one example of the gendered impacts of service withdrawal. Nevertheless, there is no doubt that both women and men as service users are significantly affected by the reduction in government and non-government services. For example, the loss of services in isolated communities creates significant hardship for residents often faced with long trips on marginal roads with little or no access to public transport. Business owners faced with closure of local bank branches must regularly travel long distances with large amounts of money in their cars. Other vulnerable members of the community face similar experiences. For example, the regionalisation of Centrelink offices means that welfare recipients in small towns are required to undertake long journeys to regional cities for Centrelink interviews or face loss of benefits. A study of Tumbarumba in New South Wales, by Wilkinson et al. (2003: 149), noted that 'a great deal of governmental policy and regulation assumes that access to a major regional town is a prerequisite in terms of fulfilling welfare obligations', creating significant difficulties for low-income local residents, who often lacked access to either adequate public transport or a reliable private vehicle. The lack of public transport and the loss of services occasionally create gendered responses. For example, Dibden and Cocklin (2003: 191–92) describe a community initiative – proposed and supported by women – to purchase a Red Cross car, which takes people to Melbourne for medical and other services; as reported to the researchers, the money for the car was raised largely by women as the male-dominated local government group did not see this as a priority funding issue.

There is some suggestion that the reduction in services has gendered impacts not only on residents as users of services, but also as workers in these services and as carers of service users. Thus, all residents, not just pregnant women, are affected by the downgrading of the small rural hospitals. However, it also has significant impacts on those local people who have lost their jobs as nurses, domestic staff and allied health providers, a large proportion of whom are women. Similarly, the loss of a bank signifies a major inconvenience for customers and businesses, but also lost are the handful of bank tellers' positions, often but not exclusively held by women. In many cases, these jobs have provided the much-needed off-farm income for women from farming families (Alston, 1995), so their loss not only signifies a loss of jobs, it also has a domino effect on the viability of farm families in the region. The movement of young women away from rural areas in search of education and employment (detailed below) also impacts on farm family and rural community viability because of the gender imbalance in the populations left behind.

The third major group of people affected by the withdrawal of services are the carers of the sick, the elderly and the disabled. Again, a gendered impact can be observed in small rural communities, predominantly because the great majority of carers are women (Alston, 2002). The loss of, or non-availability of, childcare, disability services and aged-care facilities, such as day and respite care, has a major impact on carers. The implication that people in small isolated rural communities can somehow get by with reduced services has major impacts on the lives of carers, restricting their availability for work and/or forcing them to make accommodations to facilitate their work roles.

A fourth group is also impacted by service loss – the volunteers who work to cover the services removed. There is an expectation that community members are available to perform voluntary work supporting local service infrastructure. For example, voluntary services, such as school tuckshop duty, listening to the reading of young children and assisting with sports days are built into schools' expectations of the community. In a similar fashion, aged care is augmented by such voluntary services as Meals on Wheels. These voluntary but essential rural community roles are again often filled by women (Alston, 1995). For rural men, the additional expectations on their voluntary time can be seen in their efforts to provide community infrastructure and support. For example, in the New South Wales town of Tumbarumba, studied by Wilkinson et al. (2003), many of the local men donated voluntary time and effort, and provided the earthmoving equipment needed, to upgrade the racetrack in the absence of government support.

With the withdrawal of government and non-government health, welfare and other services from small communities, there is an added expectation that the community will fill the gap through voluntary efforts. The Red Cross car in 'Tarra' is but one example of this. The study of Tumbarumba found that the withdrawal of services is placing great pressure on the community (Wilkinson et al., 2003). Those suffering mental health problems, for example, are seeking out ministers of religion or non-government organisations for assistance. Neighbourhood Centre volunteers are providing services with little or no government support or funds and few resources, and, in many cases, these volunteers are aged, drawn into the work by their strong commitment to providing for their communities in the absence of public assistance.

For women, the expectations associated with their family role and the constant community need for volunteers leads to quite complex negotiations at the gender regime level. In previous work, I have noted that working women will go to great lengths to comply with these gendered community expectations (Alston, 1995). For example, it is not unusual for women to pay someone to work in their business while they do Meals on Wheels or reading at the school. Others report that they may turn down casual teaching days because of their voluntary roles in the community (Alston, 1995). For many women and men, gendered expectations and a loss of services cost money, create inconvenience and restrict their options as service users, workers, carers and volunteers.

HUMAN CAPITAL

An examination of human capital resources in small rural areas again reveals that living in a rural community is a gendered experience. Significant changes have occurred in small inland rural communities of Australia since the 1980s, particularly those reliant on agriculture. Globalisation, low commodity prices and cost-price squeezes have made agriculture less attractive to a more mobile younger generation. Combined with easier access to higher education, new employment opportunities in urban areas and loss of employment options and services in rural communities, major population shifts have occurred in inland areas. Smailes and Hugo (2003) note that those who leave their small community for higher education are almost invariably lost to the community and are only partly replaced by the in-migration of people with higher education. As a result, they note, less than 6 per cent of the Gilbert Valley community hold Bachelors or higher degrees.

POPULATION CHANGES

Salt (1992, 1999, 2001) notes significant population shifts from smaller communities to 'sponge' cities in the regional areas and a drift of population to the coastal areas. This trend is evident in the rural communities described in Cocklin and Alston (2003), which mainly have static or declining and ageing populations. Taking a 'snapshot' of selected areas to illustrate this trend, table 8.1 indicates that all showed a decline over a ten-year period except the Gilbert Valley, where boundary changes in Census areas make these figures uncertain. It should be noted that statistics from other small inland communities show similar trends (Alston, 2002).

Table 8.1 Population changes in six case study areas

	1991	1996	2001
Guyra	4734	4262	4167
Tumbarumba	3693	3613	3551
Wellington/ Alberton ('Tarra')	6165	5722	5579
Narrogin	4661	4498	4395
Monto	1338	1293	1083
Gilbert Valley	7432	7796	8047

Source: Data from Australian Bureau of Statistics website and ABS (2002d)

Of significance for rural communities and for any discussion of gender in rural areas is the greater loss of younger women (Alston, 2002), a trend also noted in Ireland by Laoire (2001) and in Norway by Dahlstrom (1996). Young women report that they are leaving because of the lack of employment options, the need to access higher education and the need to escape the masculine culture, evident in the sporting profile, power structures and employment opportunities, that restricts their choices (Alston & Kent, 2001). Some counter in-migration is evident in small communities with, for example, single mothers and their children moving into some areas because of cheaper housing options. Similarly, in the 'Tarra' district, studied by Dibden and Cocklin (2003), there are greater numbers of older women retirees living in close proximity to health and aged services in town, or remaining in the coastal region after the death of their husbands, a trend mirrored in other areas where females dominate the over-65 age groups. However, overall there are larger numbers of men than women in all the communities listed in Tables 8.2 and 8.3. In Tumbarumba, for example, 54 per cent of the population is male compared to 46 per cent female (Wilkinson et al., 2003).

Table 8.2 15–24 year age groups in six case study areas

	1991			1996			2001		
	Male	Female	Total	Male	Female	Total	Male	Female	Total
Guyra	359	321	680	258	256	514	232	182	414
Tumbarumba	243	180	423	224	153	377	191	157	348
Wellington/ Alberton ('Tarra')	380	322	702	309	262	571	307	284	591
Narrogin	420	407	827	352	357	709	298	375	673
Monto	76	99	175	88	84	172	56	59	115
Gilbert Valley	400	351	751	349	334	683	443	361	804
Total	1878	1680	2878	1580	1446	3026	1527	1418	2945

Source: ABS (2002d)

Table 8.3 65+ age groups in six case study areas

	1991			1996			2001		
	Male	Female	Total	Male	Female	Total	Male	Female	Total
Guyra	250	268	518	281	302	583	305	314	619
Tumbarumba	238	243	481	217	238	455	257	280	537
Wellington/ Alberton ('Tarra')	376	424	800	406	444	850	426	475	901
Narrogin	231	311	542	238	341	579	229	316	545
Monto	100	121	221	116	161	277	110	159	269
Gilbert Valley	478	605	1083	520	653	1173	564	681	1245
Total	1673	1972	3645	1778	2139	3917	1891	2225	4116

Source: ABS (2002d)

As the tables demonstrate, there is a loss over time of young people (table 8.2) and a growing number of people over 65 (table 8.3). Small rural towns are becoming ageing communities, with fewer services, cared for often by ageing volunteers struggling to 'paper over the cracks'.

MENTAL HEALTH

High levels of mental illness are evident in many communities where few services are available to care for complex conditions. In Tumbarumba, for example, Wilkinson et al. (2003) found that mental health services are provided by a visiting adult worker on one half day a week, with a similar service for young people; the nearest psychiatric service consists of a visiting service to a regional centre 70 kilometres away, providing only five consultation sessions every two months. Mental health appears to have gendered manifestations, one example being higher levels of male suicide in rural communities. For women, a high level of domestic violence is one of the precipitating factors in mental illness. It is perhaps around this issue that the gendering of rural communities is most evident. High levels of domestic violence may suggest patriarchal resistance to changes in society. For women, the experience of rural community living is often coloured not only by their experiences of violence but also by a lack of support, services and options.

Few small towns have refuge accommodation, although there is usually provision made by service providers and police to take women to refuges in other communities. For some women, attempts at flight are thwarted by the fact that there may be only one source of public transport leaving the town, making it very easy for perpetrators to prevent the flight of abused women. Having to leave their community is a poor substitute for effective services and most women struggle on in silence. Other complexities of rural community life, including the fact that the local policeman may socialise with the perpetrator, may prevent women seeking help (Alston, 1997).

DIVISION OF LABOUR

Hegemonic masculinity and the gendered experiences of rural community members are also evident in the normalisation of the division of labour in rural Australian communities. Men dominate agriculture and the allied industries and businesses, such as machinery dealerships and stock and station agencies. Men also dominate in certain influential professional roles, such as medicine, law, religion and in senior local government positions. Additionally, males dominate local apprenticeships. In one town, more than 90 per cent of apprenticeships are taken by males, with all but one of the female apprenticeships being in hairdressing (Wilkinson et al., 2003).

Women are more likely to be found in lower status roles, such as nursing, administrative and customer service roles. Women are also more likely to gain casual employment in horticulture positions, such as grape pruning and nursery work (Wilkinson et al., 2003; Dibden & Cocklin, 2003). For young women who remain in rural communities, the lack of secure employment options leads to a sense of entrapment (Wilkinson et al., 2003). Professional women who move into a rural community to facilitate their husband taking on a professional role may also find the same experience awaits them. Often there is no position for the female partner to pursue her own career, resulting in women with professional qualifications taking on unskilled work in supermarkets or casual farm labour.

Rural labour market segmentation means fewer opportunities for women. The lack of employment options results in lower levels of female labour force participation. Dibden and Cocklin (2003), for example, found that labour force participation by women in 'Tarra' was 39.6 per cent, compared to the male rate of 60.5 per cent. Women in small communities note that winning one of the scarce 'female' jobs depends on whom you know (Wilkinson et al., 2003; see also Alston & Kent, 2001). Women's labour market participation is also constrained by their responsibility for domestic tasks and carework. As a consequence, they are more likely to work part-time and their employment conditions are more likely to be precarious.

As Little (1994) notes, women's participation in the labour market is as much to do with the lack of positions as with the operation of gender relations in rural households and communities. Because of the greater advantage that men gain from labour market segmentation, negotiations at the gender regime level of the family allow men to view their domestic role as 'helping' rather than participating. Women are far more likely to be responsible for the domestic labour in rural households and for the family caring roles in relation to child care, aged care and disability care (Alston, 2000). Halliday and Little (2001) note that the cultural construction of women's mothering and childcare role is central to the traditional notion of the rural idyll. Women who work must therefore fit this around their central mothering role. As a consequence, women's interaction with the marketplace is viewed as secondary and therefore the provision of childcare services is not a major consideration for those with the power to intervene.

Halliday and Little (2001) note a link between rurality, low levels of services and the expectations on mothers. For women working outside the home, their employment must necessarily fit with available child care. The resistance of women to this gender construction can be seen in the high use of informal care services (grandparents, other family members, neighbours and friends) and their efforts to seek paid employment (Alston, 2002; Halliday & Little, 2001). Yet, there is a distinct lack of contestation within the community at large about the lack of childcare services, with the result that it is women who must juggle their workloads, locate child care in a limited market and often in the absence of family, and cope with the problems associated with sick children and caring for relatives.

However, no division of labour is normal or straightforward. Rather, power and gender relations and gender identities combine with and through discursive practices in everyday life, resulting in a hierarchical gendered division of labour. Negotiations around these roles, particularly in the private sphere at the gender regime level, are ongoing and constant. Hegemonic masculinity ensures that men have a stronger negotiating position around domestic labour and therefore may make themselves unavailable for household work. Younger women, in particular, report that their partners' consent to take on a large share of household responsibility is at best grudging (Alston, 1995). It is evident that the expectation that women will be almost solely responsible for these tasks goes against young women's notions of gender identity and understanding of gender relations, often formulated in an urban environment. Their scrutiny of the discursive practices around these roles is therefore intense.

SOCIAL CAPITAL

Social capital is enhanced by the networks of trust built at rural community level. In many rural communities, not only the organisations but also the ways in which individuals interact with them are largely gendered. Many community members belong to several organisations and networks. Wilkinson et al. (2003), for example, reported that Tumbarumba, with a population of 3551, has 15 art and craft groups, 16 community support groups and 34 sports groups. Smailes and Hugo (2003) noted that the 186 households surveyed for their Gilbert Valley study recorded 66 different social organisations. They also observed that only 16 per cent of women in their Gilbert Valley sample,

compared to 25 per cent of men, did not belong to any local organisation, while some residents reported belonging to as many as 11 organisations. These figures suggest strong and deeply enmeshed community
networks in small rural communities. They also suggest that women, as
Dibden and Cocklin (2003) observed, are more likely to engage in
networks of trust and prefer more community-oriented and less
confrontational styles of community organisation.

ENVIRONMENTAL CAPITAL

The environmental impacts in rural communities of agriculture, new
industries and rural community living are well documented in the case
studies that underpin this work (Cocklin & Alston, 2003). No claim is
made that there are differential gendered impacts on the environment.
One can only note that the control of agricultural resources, land and
businesses in small rural communities is gendered. To suggest that
conditions would be different if women held more control would be to
draw a very long bow. Nevertheless, it is noteworthy that women's
organisations emerging since the 1990s have created significant space in
their agendas for discussions of the environment, chemical overuse,
negative impacts of genetically modified organisms, the health of rural
community members and ecological sustainability (Alston, 2000). With
greater access to resources and traditional sources of power, it may well
be that women would create a more focused environmental agenda.

CONCLUSION

Hegemonic masculinity dominates rural communities, shaping a
gendered experience of rural community living. Gender negotiations in
rural areas occur within the gender regime of the family and community
against a backdrop of a gender order that subordinates women. Studies
in small rural communities reveal that the experience of rural community living does indeed differ for women and men. Men are more likely
to work full-time, to hold public positions of power and to associate
with sites where masculine hegemony is reinforced. Women are more
likely to have significant responsibility for household and carework, to
work part-time, to undertake community volunteer work and to be
victims of violence.

Power in rural communities is more likely to be held by men than
by women, as are much of the wealth and resources of small communities. The fact that this goes largely unremarked is indicative of the

normalisation of hegemonic masculinity and its associated power and prestige. Masculine hegemony dominates and shapes rural communities, and sites for the creation and reinforcement of this hegemony, such as pubs and sporting fields, are commonplace. Less obvious is the fact that similar sites for women, such as craft groups, are less public, not resourced to the same extent by the community and are, therefore, only marginally effective at changing the public discourse. While rural women are increasingly inclined to scrutinise such hegemony, it is evident that the benefit accruing to men, the patriarchal dividend, ensures that men are less willing to admit their privileged position or to forego this privilege. Rural community studies reveal that masculine hegemony is highly resistant to change (see, for example, the summary in Whittenbury, 2003). Nonetheless, there is evidence of some resistance by women. The greater proportion of young women lost through the significant out-migration of young people, the seeking of higher education, and the growing reluctance of young men to enter farming and of women to marry farmers (Alston, 2002) are all signs of the erosion of the legitimation of patriarchy. Intense scrutiny of the discursive and institutional practices that shape hegemonic masculinity gives some hope that changes will occur in the gender order, allowing greater opportunities for rural women and men.

Many rural communities are communities in transition. The loss of young people, the ageing of rural populations and the withdrawal of services has placed strain on communities and has had differential impacts on women and men in their roles as users, workers, carers and volunteers. The representation of a singular rural discourse – for example, in popular political commentary – overshadows these internal differences between groups.

There is no doubt that, as Whatmore et al. (1994) note, gender remains a marginalised concept in rural community studies, despite the overwhelming evidence of its importance in shaping rural experience. Either gender is not used as an important concept or the experience of women is largely ignored. Through the inattentiveness of rural scholars to gender implications, we provide support for pervasive masculine dominance. That masculine hegemony can be so overwhelming and yet go so unremarked indicates a particular blindness to the lived experiences of many rural dwellers. It is necessary to ensure that gender moves out of the women's studies area and women's units and into mainstream government departments and research agendas. To do less is to risk continuing a process that subordinates women.

9

SOCIAL EXCLUSION IN RURAL AUSTRALIA

Margaret Alston

There is no doubt that inequalities within Australian society are increasing (Ife, 2001). What is disturbing is that inequality appears to have a postcode and that rural areas are over-represented among the most disadvantaged, defined around a number of critical indicators of inequality (Vinson, 1999). As the gap between the well-off and the poor widens (Townsend, 1996), rural people are disproportionately disadvantaged in this equation. Inter- and intra-regional disparities are changing not only the face of Australian society but also the traditional Australian ethos of egalitarianism. The causes of this change are not hard to find. Globalisation, reduced government involvement in service provision, a prioritising of the market under policies of economic fundamentalism, a reshaping of social policy initiatives around a more punitive approach to disadvantage and a decline in the welfare state are critical to the developing inequity in Australian society. While poverty is one aspect of escalating inequalities, a more thorough explanation can be found in the concept of social exclusion.

In this chapter, I consider the growing number of socially excluded people in rural areas, arguing that they are caught in a cycle of poverty and disadvantage brought on by a lack of access not only to resources but also to services that allow greater participation in society, such as education, health, welfare and employment. Townsend and Gordon (2000), in a United Kingdom context, argue that there is not just growing poverty in society but also growing social polarisation within and between countries and that the rapidity of this development has made it difficult for governments to respond appropriately. The divide between rural and urban areas demonstrates this polarisation within Australian society. Despite increasing disadvantage, the provision of services and processes that allow increased participation are more limited in rural areas, exacerbating the social exclusion of the rural poor. Moreover, a continuation of neoliberal policies of market prioritisation and reduced government involvement will increase the gulf between rural and urban Australians, unless moves are taken to redress the imbalance.

WHAT IS SOCIAL EXCLUSION?

Social exclusion refers to processes that marginalise certain groups in society and create pockets of disadvantage and poverty. It has been defined in the United Kingdom (SEU, 2001: 10) as a:

> ... shorthand term for what can happen when people or areas suffer from a combination of linked problems such as unemployment, poor skills, low incomes, poor housing, high crime, bad health and family breakdown.

It is notable that the UK definition explicitly identifies socially excluded *areas* as well as individuals, supporting the notion of geographic disadvantage put forward by Vinson (1999) in the Australian context. Demonstrating that the concept is much broader than poverty, Burchardt (2000: 388) provides the following definition:

> An individual is socially excluded if he or she does not participate to a reasonable degree over time in certain key activities of his or her society, and (a) this is for reasons beyond his or her control and (b) he or she would like to participate.

Lack of participation in society is also noted by Walker (1997: 8), who argues that 'social exclusion' is a much broader term than 'poverty' and includes not only material deprivation but also an inability to participate 'effectively in economic, social, political and cultural life', leading individuals to become alienated from mainstream society. There is no doubt that, on balance, many individuals living in rural communities are becoming socially excluded, as are rural areas themselves as a result of decreasing access to services and employment opportunities and the ageing of rural populations.

Ferge (2000) and Ruspini (2000), writing from Hungarian and Italian perspectives respectively, both note that women are over-represented among the ranks of the socially excluded and that single mothers and aged women constitute significant proportions of the socially excluded. If we examine the profiles of small rural town populations, and, in particular, the greater number of women among the rural aged, we can see the propensity for women living in rural communities to be socially excluded.

While social exclusion provides an expanded notion of deprivation, it is nonetheless useful to heed the warnings of Burchardt (2000) and Walker (1997) that the concept allows conservative politicians to avoid any discussion of poverty, its structural causes and the flawed ability of the capitalist system to provide safeguards for the poor. Further, the concept implies that social exclusion can be overcome if individuals simply adhere to certain practices that enhance inclusion, thus endorsing a 'blame the victim' approach to deprivation, while structural influences are not acknowledged. Nonetheless, the term has become commonplace in the United Kingdom, where a special Social Exclusion Unit has been established in the Deputy Prime Minister's Office aimed at addressing the disadvantage suffered by certain identified groups, including significant numbers of rural people. Work conducted under the auspice of this unit identifies key risks associated with social exclusion as being low levels of education, low incomes, unemployment, poor housing, health problems and family breakdown (SEU, 2001), which can be concentrated in identifiable areas. Burchardt (2000) also argues that geographical concentration plays a part in social exclusion and this is supported by United Kingdom work that reveals increasing social exclusion in rural areas, a position also endorsed by research in the European Union revealing significant rural disadvantage (Ray, 2000).

If we understand social exclusion to be an inability to participate effectively in society for reasons beyond the control of the individual, and rural disadvantage has been identified as a key factor, then this places many rural Australians in the category of the socially excluded. When we examine the changing dynamics in rural Australia, particularly in inland areas, we can see that there are significant groups of rural people at risk of social exclusion: the unemployed, single mothers, the aged, the 'working poor', young people, people from non-English speaking backgrounds and Indigenous Australians. Yet it is significant that 'social exclusion' has not entered the discourse of the present conservative Australian federal government, although the growing number of socially excluded rural Australians is tacitly acknowledged. Speaking in 1999, the Deputy Prime Minister and leader of the National Party, John Anderson (1999), noted that the sense of alienation and of being left behind in rural Australia was deep and palpable. The failure by the Australian government to name 'social exclusion' appears to result from a slavish pursuit of market primacy, economic rationalism and minimal government intervention. It is also clear that the federal government is reluctant to pursue an agenda championing social justice, preferring to endorse individualism over collective actions. In the process, what is evolving is the replacement of the 'social rights of the entitled citizen of the welfare state ... by the social obligations of the dutiful citizen' (Powell, 2001: 16), creating, as Cox and Caldwell (2000: 71) argue, 'mutual mistrust and coercive reciprocity' and further alienation among the most vulnerable.

While the concept of social exclusion is readily accepted in the United Kingdom, Levitas (2000) argues that discourses around social exclusion are often contradictory and confusing. She suggests that these can be broken into three competing approaches. The first and most simplistic is the redistributive approach that implies that social exclusion is a consequence of poverty and is limited to a lack of money. The second is the social integration discourse, which suggests that labour force attachment is the key to social exclusion and that those outside the labour market are at a greater risk of becoming excluded. The third is the moral underclass discourse, a discourse that suggests moral and cultural causes of poverty and implies that the socially excluded pose a threat to the moral order. Policy initiatives of the Howard government suggest a combination of these discourses. Strong emphasis is placed on participation in the workforce, while a moral undertone implies that certain groups of the impoverished, such as single mothers and young people, need careful management and potentially harsh punishments.

GLOBALISATION AND NEOLIBERALISM

One of the critical factors leading to inequalities among regions across the world has been globalisation, the development of a global marketplace and the erosion of national boundaries. The OECD defines globalisation as:

> ... increased movement of tangible and intangible goods and services, including ownership rights, via trade and investment, and often of people via migration. It can be and is often facilitated by a

lowering of government impediments to that movement, and/or by technological progress, notably in transport and communications. (Oman, 1996: 6)

The breaking down of national protection has exposed some areas to the perils of competition. Critical to success in a global environment, as the OECD argues, is access to communications technology and transport infrastructure (Oman, 1996), allowing the facilitation of international networks (Ife, 2001). For rural areas, such access is particularly difficult, as the combination of market forces and a lack of critical mass create serious service shortages.

While globalisation has impacted on rural areas, it is the championing of neoliberal ideologies within Australia and elsewhere that has marginalised vulnerable people. These ideologies, favouring minimal government involvement in the marketplace and market dominance over social policy, promote individual freedom while downplaying collective action (Gray & Lawrence, 2001a; Dominelli, 2002). Australia's version of neoliberalism has done much to foster the growth of corporate bodies, while at the same time making it more difficult for disadvantaged groups of Australians to receive income support. In fact, it would appear that the attention to profit generation has become linked with a lack of attention to citizenship rights (Dominelli, 2002). As Tonts demonstrates in chapter 11, previous policies supporting the development of inland areas through a process of cross-subsidisation no longer apply, leaving many inland regions lagging (see also Tonts, 2000). Policies such as the National Competition Policy have been particularly severe, robbing inland rural areas of services and infrastructure. In many such regions, transport is inadequate and/or expensive and communications infrastructure, such as Internet and broadband access, and mobile telephone coverage, is deficient.

In Australia, evidence of New Right discourses around policy are not hard to find. The discourse that shapes policy has changed perceptibly in Australia over the last two decades, from one resonant with ideas of citizenship rights, social justice and equity towards economic fundamentalist positions that champion profit, the market, managerialism, privatisation of services, quality assurance, accountability and mutual obligation (Alston & McKinnon, 2001). There has also been a commensurate shift away from the notion of community obligation to provide for the needy or disadvantaged towards a position where individuals and communities are held responsible for their own welfare and have a mutual obligation to society in exchange for benefits. As a result, services are rated on economic efficiency rather than need and, without a critical mass to provide a secure financial base, many inland rural areas have suffered.

There is also ample evidence of a more punitive attitude to the most vulnerable in Australian society, not only at the government policy level but also from the community itself. The rise of the New Right and a move to the right in the community more generally has strengthened this position. The federal government's welfare reform package has made it mandatory for people to socially and economically participate in society in return for benefits (DFCS, 2000). Thus, the most vulnerable must meet certain requirements in return for benefits. Those who do not are subject to 'breaching' or the cutting or total withdrawal of benefits. The Australian Council of Social Services' study of breaching in the period from June 2000 to June 2001 reveals that 350 000 unemployed had their benefits cut, a rise of 189 per cent, and 13 500 had benefits withdrawn altogether. Over-represented in this group were the mentally ill, the disabled and the homeless (ACOSS, 2001). There is also some evidence that rural people have been particularly affected by more punitive policies surrounding welfare entitlement (Alston & Kent, 2001).

Meanwhile, globalisation has aided the restructuring of the workplace, with a decline in the number of full-time jobs and a growth in part-time casual positions. Across Australia, and not just in rural areas, the proportion of Australians in full-time work was 20 per cent lower in 1998 than it was in 1973 and, during the same period, the proportion in part-time work trebled (Sheehan & Tegart, 1998). Workplace conditions are also being undermined with a growing emphasis on casualisation, and a decline in security of employment. In rural areas, unemployment is higher and more prolonged, and job opportunities are more limited, poorly paid and, in many cases, seasonal (Cheers, 1998). Workplace restructuring has had a critical impact in rural areas, with a loss of permanent positions and an erosion of workplace conditions. This, together with the withdrawal of many thousands of government positions from rural areas (Wahlquist, 1997), has increased the numbers of rural Australians in danger of becoming socially excluded through long-term unemployment.

For rural people and communities, the self-help philosophy fostered by governments is evident in government funding and programs. For example, the Regional Solutions Programme is represented, in its guidelines, as 'an innovative Federal government initiative helping regional and rural communities find local solutions to local challenges' (DOTRS, 2002). The overarching philosophy for regional programs is spelled out in the policy document. Here the approach is noted as 'a partnership that fosters the development of self-reliant regions' through 'regional communities managing change and leading their own development', with the 'Federal Government supporting new ideas, self-reliance and achievement' (DOTRS, 2001: 4).

The expectation of governments that disadvantaged individuals and communities can be self-reliant supports a position of limited government involvement and reduced funding. In an ideal world, individuals and communities may well become self-sustaining. However, in a situation where market dominance is supreme and minimal government intervention is favoured, the poor and the more vulnerable communities are at serious risk and the number of socially excluded people escalates. As the following 'snapshot' shows, Australia's rural communities have been particularly vulnerable to globalising forces and rural people feature prominently in the ranks of the socially excluded.

POPULATION CHANGES

Since the 1980s, significant population movement has occurred from the inland areas to the capital cities and the coast (Salt, 2001). Many inland communities are experiencing population loss or stagnant growth. As Salt (2001) notes, changes are not uniform and population shifts to the coastal regions of the eastern seaboard, for example, have seen significant growth in these areas. Of significance to this paper is the work of Ian Burnley and Peter Murphy (2003), who observe that the trend from metropolitan to non-metropolitan areas is driven by a number of factors, including the movement of people in search of a 'seachange'. Of note is Murphy's (2002: 7) contention that among those seeking an enhanced lifestyle are also many 'forced relocators', those forced out of the cities by financial necessity. These forced relocators are generally low-income earners on income support and many are unemployed or in single-parent households. Murphy (2002) notes that, over a recent 12-month period, 11 000 unemployed people and 5000 single-parent households moved from metropolitan to non-metropolitan areas. This movement of forced relocators, together with aged people seeking an enhanced lifestyle, is adding pressure to local government areas for service support. While most gravitate to coastal areas, there are also many people on income support moving to inland rural areas, exacerbating issues of service decline in these areas (see, for example, Alston & Kent, 2001).

Nonetheless, many inland rural areas are experiencing population decline and the most likely to suffer population losses are those areas reliant on broadacre farming (ABARE, 1999). There is some evidence of growth in inland Australia, but the rates are slower than for urban areas. Garnaut et al. (2001), for example, report a 6.5 per cent growth in inland areas from 1986 to 1996, compared to 13.7 per cent in capital cities and 28.4 per cent in other metropolitan areas. The slower or negative growth inland is the consequence of a decline in the relative profitability of agriculture coupled with other effects of globalisation and significant periods of drought. Of particular concern is the out-migration of young people seeking work and higher education opportunities (Alston & Kent, 2001), and of professionals made redundant through the closure of rural services. Counterbalancing this in some areas is the movement of forced relocators.

The loss of young people accounts for a significant slice of rural migrants to urban areas. These young people report a strong need to leave rural communities in search of better opportunities (Alston & Kent, 2001). Meanwhile, as Laoire (2001) notes for those young people remaining, 'staying behind' presents significant challenges. Those who stay behind are particularly vulnerable to social exclusion, their alienation evident in the high number of young male suicides in rural areas (Bourke, 2001a). Recent research (Alston, 2002; Alston & Kent, 2001) reveals that young rural men dominate the ranks of those young people remaining in rural areas. This results from the more ready availability of apprenticeships in skilled areas usually taken up by men and the greater likelihood that young men will have jobs on farms and prospects of farm inheritance. For young people without these prospects, or the education and skills to obtain urban employment, their alienation is driven by a lack of employment opportunities, declining populations and communities, a loss of peers and a lack of options. The loss of local employment opportunities, including entry-level positions in government services and non-government services such as banks, has meant training sites and career positions are virtually non-existent in small communities.

What is clear is that the 'flight' from rural areas is significantly gendered. There are several reasons for young women leaving rural communities in greater numbers and these include the practice of patrilineal inheritance in agriculture and small rural businesses, and the male monopoly on apprenticeship positions: in the New South Wales town of Tumbarumba, for example, research revealed that 90 per cent of apprenticeships are taken up by young men (Wilkinson et al., 2003). Young women also note the 'macho culture' in country towns, including the expectation that women will be subordinate, the pub culture and violence against women as reasons for leaving (Alston & Kent, 2001).

The loss of young people and professionals leaves behind ageing populations in small rural communities. The ageing of the population is accelerated in some areas by the in-migration of retirees. Research suggests that the proportion of people aged over 65 in the year 2026 will vary from 18 per cent in capital cities to 25 per cent in rural communities (Birrell, 2000) and, if population profiles remain as they are now, a majority of the rural aged will be female.

POVERTY

The link between economic deprivation and social exclusion has been noted by Townsend and Gordon (2000) and Walker and Walker (1997). Without the ability to purchase services, including education, health, welfare and transport services, those on low incomes are unable to participate effectively in society and thus become socially excluded. Vinson (1999: 20) notes the effects of 'cumulative disadvantage', commensurate on a number of indicators occurring within regions. Further, Vinson's (1999) work in disadvantaged regions indicates that rural areas are significantly over-represented among the most disadvantaged. What we do know is that poverty is increasing in Australia's rural areas and that an increase in social exclusion can be assumed. The drift into these areas of forced relocators on income support increases the areas' poverty profile. Research over several years reveals growing numbers of rural poor (see, for example, Cox & Foster, 1990, and Lawrence, 1995). Recent research indicates that 25 per cent of households in inland New South Wales regions live on incomes of less than $10 400 (Garnaut et al., 2001); and non-metropolitan household incomes declined from 1981 to 1996 relative to the national average (Productivity Commission, 1999). That there are growing numbers of socially excluded rural Australians would appear incontestable.

EMPLOYMENT

Employment is a critical factor in enabling vibrant rural communities and in allowing people to participate effectively. A feature of the employment landscape in rural areas has been the cutbacks in government services such as post offices, courthouses, schools and hospitals. Many non-government services such as banks have followed suit, leaving many communities not only without services but also without important employment opportunities. The loss of skilled positions and the precariousness of local unskilled employment opportunities create an uncertain future for many communities. While employment opportunities in agriculture were the main areas of decline in inland areas between 1986 and 1996, falling by 15 per cent, employment in electricity, gas and water services declined by 50 per cent, in government administration and defence by 11 per cent, in communications by 12 per cent and in transport by 14 per cent (Garnaut et al., 2001). For young people, this has meant the loss of employment options, training sites, and agencies with clear career paths, and a dearth of satisfying work options. In some rural communities, unemployment is up to 40 per cent (DPIE & CRSR, 1997), while in some Indigenous communities it is as high as 95 per cent (Wyn et al., 1998).

As Laoire (2001) notes, employment is central to male identity and its loss has had a significant effect on rural men. The loss of large numbers of previously 'male' jobs in areas such as agriculture, telecommunications, the railways, steel works and manufacturing suggest that many unskilled rural males are among the socially excluded. Additionally, while the loss of unskilled employment is evident, there is a new class of self-employed poor, populated by farmers and small-business owners struggling with declining terms of trade and the effects of drought (Alston & Kent, 2004). Women in rural communities, particularly those in unskilled positions, have always faced rigid labour market segmentation. Employment opportunities for women have been precarious, low paid and insecure. It is arguable that rural women's attachment to the labour force remains unstable because of declining employment opportunities and the growth of casualisation. Meanwhile, many small communities report extraordinary difficulties in attracting skilled staff to positions such as shire engineers and town planners, allied health workers, solicitors, doctors and nurses (Alston et al., 2001b).

The reduction in services and accompanying loss of professionals, and the failure to attract skilled staff, has also created a vacuum in leadership. This loss is felt keenly in country towns where there are now considerable expectations on communities to help themselves. In the past, local teachers, bank managers, accountants and doctors could be expected to provide leadership for community development. With the loss of government positions in education, health, welfare and agriculture, and the downgrading or closing of non-government services such as banks, communities have been denuded of professionals and of their de facto leaders, leaving growing numbers of disadvantaged rural people.

Indigenous Australians experience much higher rates of unemployment, officially three times the rate of non-Indigenous Australians, and are less likely to find employment in rural and remote areas than in urban areas (ABS, 2003d: 4). Some 17 800 Indigenous people, mostly in remote or very remote areas, work in Community Development Employment Projects (CDEP), a scheme developed to provide employment for Indigenous workers (ABS, 2003d) as an alternative to social security benefits. This scheme has been criticised by Langton (2002: 11), who states that: 'The scheme is widely regarded by informed Aboriginal leaders as the principal poverty trap for Aboriginal individuals, families and communities.'

HEALTH

The health status of rural Australians is poorer by comparison with urban Australians. Rural Australians suffer higher morbidity and mortality rates, higher numbers of psychiatric disorders and a greater incidence of heart disease, stroke and respiratory disease (DPIE & CRSR, 1997). On many health indicators, Indigenous Australians fare particularly poorly, making them among the most vulnerable of rural Australians. They are more likely to suffer from heart disease, respiratory diseases, infectious diseases and injury than other Australians and have death rates up to five times higher than the general population (ABS, 2003c: 4). Despite the evident need for greater access to medical facilities that these statistics reveal, there is a shortage of at least 500 medical practitioners in rural areas (Harrison, 1997), and there are fewer services, longer waiting times and longer distances to travel for specialist attention, as well as lower levels of bulk-billing (Australian Institute of Health and Welfare, 2002). Further, there are very few rural communities with obstetric services that allow women to give birth in small country towns (see, for example, Stayner, 2003), creating increased hardship for rural women and a deterrent for potential newcomers.

EDUCATIONAL ACCESS

The link between education and social exclusion is well understood. Shucksmith (2000), for example, notes that education is one of the critical factors that facilitate people's access to the global marketplace. Those with lower levels of education are limited to local labour markets, severely restricting their life chances. In further research, Hobcraft (2001) notes that educational level is significantly implicated in adult social exclusion – a position also noted by the UK Social Exclusion Unit (SEU, 2001). It is therefore important to ensure that young people's educational access is not hindered in any way and is actively facilitated. Unfortunately, for rural Australians at least, there are significant barriers and rural young people are therefore more likely to join the ranks of the socially excluded. Despite rural Australians constituting 28 per cent of the Australian population, the proportion of young rural people attending tertiary institutions declined from 18.7 per cent in 1992 to 17.4 per cent in 1997, despite the fact that the total number of young people going on to tertiary education increased by 25 per cent during the 1990s (HREOC, 1999). One of the main causes of the decline in the proportion of rural students accessing tertiary education has been the lack of eligibility of many rural young people to tertiary allowances (AUSTUDY and Youth Allowance) (Alston et al., 2001b).

School retention rates are also lower in rural areas, as school children are more likely to drop out before they complete high school. For example, the Human Rights and Equal Opportunity Commisson (HREOC, 1999) notes that, in Western Australia, the dropout rates vary from 25 per cent in the capital city to between 50 per cent and 75 per cent in rural and remote areas, with much higher rates for Indigenous Australians. In some schools, the numbers completing high school are as low as one-sixth of the numbers in metropolitan schools (HREOC, 2000a). Overall, Indigenous students are only half as likely as non-Indigenous students to complete high school (ABS, 2003d: 2).

SOCIALLY EXCLUDED RURAL AUSTRALIANS

Using Burchardt's (2000) definition of social exclusion outlined above, it is clear that there are growing numbers of rural Australians who are socially excluded, unable to participate fully in key activities through reasons beyond their control and shut out of economic, social, political and cultural life of Australian society. These include the aged, young people, Indigenous Australians, single mothers, and people locked out of full engagement in the labour market, such as the unemployed and the under-employed. Further, for non-English speaking background rural Australians, social exclusion can be complicated by a lack of education, language difficulties, isolation and cultural practices

such as a reluctance to allow young women to leave home for education purposes (Alston, 2000; Alston et al., 2001b; Wilkinson et al.,
2003). To put this into perspective, 22 per cent of women and 23 per
cent of men in Australia were born overseas and of these 13 per cent
of women and 14 per cent of men are from non-English speaking backgrounds (ABS & OSW, 1995) and many live in rural communities. For
Indigenous Australians, the situation is further complicated by racism,
problems of access to education, health problems and the critical
importance of family, which leads to a greater reluctance on the part of
young people to leave home (Alston, 2000; Alston et al., 2001b).
Indigenous Australians constitute 2.4 per cent of the population, with
proportions of the resident population in rural and remote areas higher
than in urban areas (ABS, 2002c). What is most disturbing is that
Indigenous Australians suffer higher morbidity and mortality rates, die
younger, are hospitalised at much higher rates (ABS, 2003c), have
lower educational completion rates, higher unemployment rates and
lower household incomes (ABS, 2003d). They are, without a doubt,
among the most disadvantaged of rural Australians and at greater risk
of being socially excluded.

Critical to the number of rural Australians at risk is the lack of access
to services and opportunities to allow full participation. Increasing
poverty, declining full-time employment and reduced access to health,
welfare and education services limit the life chances of rural Australians.
Added to this is the changing policy focus to economic fundamentalism
and away from social justice initiatives. Nonetheless, it is important to
note that not all rural Australians are socially excluded and that rural
areas and people are characterised by diversity. Reimer and Aipedale
(2000), writing about Canada, note that not all rural people are socially
disadvantaged by globalisation. In fact, they define three groups or
'three rural Canadas' as:

- the bigger farms and businesses linked to the global marketplace and
  making significant profits – in Australia there are many such businesses,
  particularly in the newer industries such as cotton
- the middle group of active citizens, usually families with at least one
  member in full-time employment and contributing most to the lives of
  their communities
- the significantly disadvantaged marginal groups outside the market economy – the aged, the young, the unemployed, single mothers and
  Indigenous people

These distinctions have applicability in the Australian context as well. In Australia, globalisation has also created winners and losers in rural and urban areas. The problem for rural people in the third group is their precarious attachment to the labour market, their lack of access to services, and the deficiencies in infrastructure and levels of government support. These rural Australians are in danger of becoming socially excluded and alienated from society. Further, under policies shaped by economic fundamentalism, they are in danger of being blamed for their circumstances and expected to be self-reliant and self-sustaining.

CONCLUSION

Social exclusion is a useful concept for describing the growing numbers of rural Australians locked out of participatory processes. These Australians require policy initiatives that enhance participation and provide greater access to services. The current focus of neoliberal policies does not address the growing inequities in rural society. A more concerted approach requires attention to service access, greater access to health and education services, determined attention to the task of linking young people to education and employment (and enacting positive discrimination practices to achieve this), provision of services to the rural aged, supportive services for single mothers, and attention to access issues for Indigenous Australians and those from a non-English speaking background. The current punitive approach to the vulnerable is not helpful and the call for individuals and communities to be self-reliant is often unrealistic. Creative policies are needed that will link people more actively into the lives of their communities and society. Sustained attention to addressing social exclusion is necessary if Australia is to reassert its dwindling egalitarian tradition.

10

NATURAL CAPITAL AND THE SUSTAINABILITY OF RURAL COMMUNITIES

Chris Cocklin

In a controversial article in *Nature*, Costanza et al. (1997) estimated the monetary value of the world's 'ecosystem services and natural capital' to have an average annual value of $33 trillion. Irrespective of the validity of the calculations,[1] the importance of the article resides in the fact that it drew widespread attention to the immense material value of the natural environment; indeed, the estimate is far greater than the sum of all the world's gross national products. From around that time, the notion of 'natural capital' has gained substantial currency, particularly in the context of discussions about sustainability.

According to Berkes and Folke (1993; see also Pretty, 1999; van der Perk & de Groot, 2000), natural capital has three components:

- the non-renewable resources that are extracted from ecosystems
- renewable resources, which are produced and maintained by ecosystem processes
- environmental services – such as the climate, soil formation, nutrient cycling, waste assimilation – the products of ecosystem functioning

For rural commodity producers, the perennially difficult act of balance is to at once draw upon natural capital and ecosystem services, while at the same time being good stewards of the environment, thereby ensuring that their farms are sustainable – ecologically and economically. In many cases, this balance has simply not been achieved and the result has been environmental degradation. This ultimately has implications for the sustainability of individual farms, but also for the rural communities of which they are a part.

A central argument of this chapter is that the relationship between natural capital and primary production is a basic one, but influenced by many factors. These influences can draw land managers in quite different directions; for example, governments implement regulations in support of improved environmental outcomes, while market pressures can lead to increased resource exploitation. If we fail to understand the complex array of influences, we are unlikely to succeed in forging a more complementary relationship between production and the environment. In this chapter, I explore the following influences on the relationships between land-based primary production and 'natural capital':

- environmental variability and change
- public attitudes and community expectations
- markets
- demographics
- technology and innovation
- governance
- land use change and ownership

In the final section of the chapter, I provide some consideration of the relationships between natural capital and social capital, and how these two can be brought together in support of sustainability in the context of rural communities. At risk of perpetuating the conflation of 'rural' with 'agriculture' (see Scott et al., 2000), the discussion focuses principally on these issues in the context of farming.

FARMING, NATURAL CAPITAL AND SUSTAINABILITY

ENVIRONMENTAL VARIABILITY AND CHANGE

Agriculture depends on natural capital in a much more fundamental way than many other forms of production. This means that the impact of farming on the environment can be considerable and, in turn, that environmental variability and change can impact significantly upon farming. Agriculture is an excellent exemplar of what has been called environmental degradation through the 'tyranny of small decisions'

(Kahn, 1966; Odum, 1982) – 'decisions that were never consciously made, but simply resulted from a series of small decisions' (Odum, 1982: 728). In other words, individual farmers rarely set out to diminish the quality of the environment, and generally the actions of each have a relatively minor impact in themselves, but cumulatively their 'decisions' impact in significant ways upon the environment (see also Cocklin, 1993; Spaling & Smit, 1995).

Recent assessments of the state of Australia's environment, particularly its land and water resources, paint an arresting picture of decline (see, for example, Australian State of the Environment Committee, 2001; Conacher & Conacher, 2000; Yencken & Wilkinson, 2000). The Australian National Land and Water Resources Audit in 2000 (NLWRA, 2001a) found widespread excesses of nutrients, turbidity and salinity in river basins. In terms of water use, the audit revealed that both surface and groundwater resources are at a high level of development and approaching, or beyond, sustainable extraction limits (NLWRA, 2001a). A dryland salinity assessment (NLWRA, 2001b) indicated that approximately 5.7 million hectares are within regions at risk or affected by dryland salinity and that within 50 years this might increase to more than 17 million hectares. The audit suggests that up to 20 000 kilometres of streams could be significantly salinity-affected by 2050 and that about 630 000 hectares of native vegetation and associated ecosystems are at risk; this could increase to two million hectares over the next 50 years. The impact of salinity on farm values and community resources is pervasive.

> On farms, salinity reduces production, income and the capital value of land. Salinity damages infrastructure, salinises water storage, causes loss of farm flora and fauna and the loss of shelter and shade. These effects are magnified at the regional level. Salinity can, and is having a substantial impact on public resources such as water supplies, roads, buildings and biodiversity. (NLWRA, 2001b)

Environmental variabilities also have widespread effects on ecosystems, natural resources, and primary production. Topical at the present time is the dry period that has affected large parts of south-eastern Australia over the past four to five years, culminating in the drought of 2002–03. Potentially more troubling in the long term is the spectre of systematic changes in the average temperature associated with human-induced climate change and the associated climatic effects. These effects include increased heat stress in people, animals and ecosystems; increased risk of bushfires; higher energy demands; and increased demands on water supply systems (CSIRO, 2001).

Gray and Lawrence (2001a) explain the pervasive and alarming degradation of the Australian environment by agriculture with reference to a range of historical and contemporary factors. The origins of ecological unsustainability, they suggest, lie in the application of European farming techniques in an environment to which they are inherently unsuited. The pursuit of productivist agriculture has privileged short-term economic returns over the maintenance of the long-term productive capacity of the land, while in more recent times market liberalisation, the cost-prize squeeze, and the increasing influences of multinational agrifood companies have further underscored longstanding patterns of environmental degradation:

> If the present policies deal with symptoms rather than with causes and if, through deregulation in the context of globalisation, they act to devalue family farm-based agriculture and 'force' producers to act in individually rational ways which are socially irrational, it is unlikely that we will ever move from the current unsustainable trajectories. (Gray & Lawrence, 2001a: 155)

Significantly, it is not only the environment that suffers – literally millions of dollars of production are lost each year, impacting adversely on individual farmers and their communities.

PUBLIC ATTITUDES AND COMMUNITY EXPECTATIONS

Australian Bureau of Statistics surveys indicate that, while there has been a decline in the level of concern among the Australian population about the environment since 1992, the majority of Australians still rate environmental issues as important (Lothian, 2002). Indeed, between 60 and 70 per cent of Australians have consistently reported environmental concerns over the past decade. The environment ranks behind health, crime, education and employment in terms of the percentage of people who rate it as *the most important* social issue but, even so, approximately 10 per cent of Australians accord it this level of significance.[2]

Notwithstanding the fickleness of attitudinal surveys, there is enough evidence in the surveys that have been conducted over the last decade or so to confirm that the environment is an important issue to Australians generally. Voting behaviour provides further insights into the public mood.

A feature of recent electoral behaviour is an apparent new leaning towards environmental issues. This was revealed in the 2001 Western Australian state election, in which the new Liberals for Forests party took 1.7 per cent of the total vote, despite running in only eight seats. Labor, which won the election, campaigned partly on the forest management issue. Most significantly, though, the Greens hold the balance of power in the upper house.

Louise Dodson, Chief Political Correspondent for *The Age* newspaper, speculated that: 'If the environment vote continues to strengthen, it could have important effects on national politics' (Dodson, 2001). Arising out of the Victorian Liberal Party conference in March 2001, a new lobby group, Liberals for the Environment, was formed, leading to another media comment (Miller, 2001) that:

> It is a sign of internal frustration over economic agendas dominating policy at the expense of other values such as environmental protection. It also indicates concern that traditional Liberal voters in Victoria may be willing to vote against the party on environmental issues.

Subsequently, the 2002 Victorian state election further confirmed the increased profile of environmental issues in politics. *The Age* newspaper observed: 'If there's one difference between the 1999 election and this year's, it is that there is now actually a debate on environmental policy' (Fyfe, 2002).

What is the significance of these attitudinal surveys and electoral outcomes in terms of the relations between primary producers and natural capital? There is no doubt that the public at large is concerned about the environment. This translates into expectations about environmental quality. In a recent report on landholders' perspectives on sustainable land management (Cocklin et al., 2003: 15), a farmer was quoted as saying:

> You've got lots of people saying you should do this, you should plant trees, you shouldn't plant trees or native pastures or ... chemicals we can or cannot use and a lot of the time it's generated from ... the general public in Melbourne or the cities ... and then we're expected to run our properties that way.

In a similar vein, a Victorian catchment management authority observed: 'Global and local community attitudes towards environmental issues have changed in the direction of greater responsibility' (Goulburn Broken CMA, 1999: 14). The fact that the environment has achieved positive electoral outcomes in recent contests adds to the potency of these demands.

MARKETS

Market trends are relevant to natural resource management (NRM) in a variety of ways. Most directly, the net returns that producers receive influence decisions about both inputs (for example, land, labour, water, fertiliser, herbicides, management and innovations) and outputs (the types of commodities produced and the nature of the economic enterprise generally). In the short term, market trends are more likely to influence decisions about inputs, whereas decisions about outputs are typically medium to longer term, given the associated costs of change.

It has been widely documented that the general trend in agricultural commodity prices has been downward. In the mid-1950s, the ratio of the prices received by farmers to the costs of production was four times higher than it is today (ABARE, 1997, 2000). The trend in commodity prices has been a relatively long-term one:

> While nominal (undeflated) agricultural commodity prices ... have risen, 'real' prices (market prices deflated by the Consumer Price Index) have declined. Real per-unit prices of inputs have also fallen, but at a slower rate. As a result, the real net income per farm in Australia has fallen steadily over the past 40 years and is now only about 60 per cent of its level in the 1950s. (Commonwealth of Australia, 1996: 6–36)

Significantly, this report goes on to note that this cost-prize squeeze has 'placed enormous pressure on farmers to increase their productivity by applying new technologies that will make production more efficient *while not necessarily accounting for any adverse changes in resource condition*' (Commonwealth of Australia, 1996: 6–36; emphasis added). Land and water resources come under increasing pressure as a consequence of the cost-prize squeeze. It has been noted, for example, that: 'Farmers are not generating a financial return sufficient to encourage allocation of a proportion of revenues towards environmental amelioration/sustainable land management' (Gippsland Development Ltd, 2001: 17).

The potential inconsistencies between industry trends and social expectations in relation to the environment are presented in box 10.1. This points to the pressures on dairy farmers for increased productivity and sets these alongside community expectations in terms of environmental management.

Box 10.1 Dairy industry trends and expectations

Dairy industry trends	Community pressures and expectations

• Intensification:
  – increased herd sizes and stocking rates
  – increased use of irrigation water and fertilisers
  – increased use of purchased feed and feed pads
  – increased use of employed labour
• Future uncertainty following deregulation
• Increased recognition of environmental impacts
	• Increased duty of care by agriculture and other industries to ensure:
	  – food safety and production system ethics
	  – improved water quality
	  – salinity protection
	  – environmental flows
	  – biodiversity

Source: DRDC/NLWRA (2001)

The prospective tensions outlined here have been exacerbated in some regions following the deregulation of the industry in mid-2000 (see Cocklin & Dibden, 2002a, 2002b). On the basis of interviews with dairy farmers in the community of Monto, Queensland, Herbert-Cheshire and Lawrence (2003: 17) noted, for example, that:

> ... expanding production is the only way to survive deregulation, although most Monto dairy farmers have been unable to follow this course of action because they are already operating at optimum stock levels for the size of their farms, or the amount of water that is currently available.

They also reported that:

> Damage to pasture has occurred as a direct result of the contemporary challenges facing Monto's rural industries. This was pointed out by one farmer, who explained that in an attempt to remain viable, dairy farmers and graziers were maximising stock levels and allowing their pastures to be completely eaten out. (Herbert-Cheshire & Lawrence, 2003: 20)

There are other market trends that bear directly on production decisions, and that also have prospective implications for natural capital. Consumer attitudes towards the environment are playing a more pronounced part in purchasing decisions. According to Gippsland Development Ltd (2001), the markets for organic food products are increasing in some countries at rates of 20–40 per cent per annum; the domestic market is increasing at approximately 25 per cent per annum. In concert with this trend, the giant UK food retailer Sainsbury's has a goal to: 'Influence our suppliers to reduce their direct environmental impacts and improve the environmental quality of own-brand products through more sustainable sourcing'.[3] Recently, the Australian forestry company (and hardware retailer) Bunnings was under threat of losing an export contract with the UK firm Railtrack for railway sleepers because the timber did not meet Forest Stewardship Council (FSC) certification requirements (Yencken & Wilkinson, 2000). On the issue of consumer demands, the Goulburn Broken CMA (1999: 14) noted that:

> Community attitudes affect industries and ultimately individual primary producers in that the community – the consumers and community shareholders – are demanding production of environmentally and socially friendly goods. Hence, market forces are playing a large role in determining the actions of producers of goods, including land managers.

Despite the optimism that this might engender, the fact remains that consumers are motivated largely by prices. If the prices of agricultural commodities, domestically and internationally, do not internalise the costs of environmental degradation, there will be little market incentive for consumers to shift their preferences towards food and fibre products that are less environmentally damaging. As one farmer, participating in a land stewardship study, put it:

> Unless the connection between consumers and land degradation is made clear, and the public benefits of sustainability at the farm level more tangible, then the resource will decline, i.e., consumers want cheap produce, but the environmental cost is not necessarily built in. (Cocklin et al., 2003: 16)

DEMOGRAPHICS

As chapters 4 and 5 reveal, the demography of Australia's rural communities is changing. These socio-demographic trends have wide-ranging implications for environmental management. For example, a declining farm population and the prospect of continuing financial stress among some rural producers raise questions about the ability of farmers and rural communities to participate in natural resource management activities. Barr et al. (2000) point to the fact that, in parts of the Murray-Darling Basin, the median farmer age is approaching 60 years. An ageing farm population has implications in its own right for NRM practices, but it is also evidence of less intergenerational transfer of properties. Barr et al. (2000: 44) show that farmers who do not plan to pass their properties on to a child have been significantly less likely to adopt new NRM practices. Similarly, Gray et al. (2000) found that the presence of an heir already working on the farm was the only factor to show a strong correlation with sustainable farming practices. They argue that the likelihood of young people wishing to remain in rural areas and hence join their parents on the family farm is increased if community ties are strong. The absence of opportunity to hand on properties is closely linked to decreased profitability of farming enterprises. A report on the development of social capability in support of NRM observed that the 'overall trends indicate that in many places social capability will [have to] be generated in the context of a declining and ageing population, with fewer people employed directly in agriculture' (MRAP, 2001: 42).

TECHNOLOGY AND INNOVATION

Technology (interpreted broadly) constitutes both a threat and an opportunity in terms of the sustainable management of natural capital. On the one hand, new technologies pose considerable risks to the quality of the environment and to communities based on primary production. Technological innovations, new scientific understanding and improved management techniques (for example, farm management plans) also hold the promise of substantially improved environmental practice. The extent to which farmers embrace innovations in management and technology in respect of water use and land management will be one of the important variables determining the rate of environmental degradation in the future. However, cost-effectiveness is as likely as environmental concern to be the impetus for change to better irrigation and other resource management strategies. As Cullen (2001: 2) observed in respect of the funding and uptake of research: 'The focus tends to be on production aspects rather than avoiding degradation, achieving long term sustainability or minimising externalities, which many users to do not want to think about.'

Genetically modified organisms (GMOs) are a development of particular significance. Consumers in many western nations have resisted the introduction of GMOs on the basis of concerns about the implications in terms of human health and the environment. Significant debate continues within the science and policy communities about these implications. In some countries, policy decisions have been responsive to consumer concerns. In Australia, though, both the Commonwealth and several state governments have taken the view that GMOs are essential to future competitiveness in both science and agriculture (Hain et al., 2002). In terms of natural capital, benefits of GMOs are said to include higher productivity and a reduced need for chemical inputs. On the other hand, critics refer to the risks of cross-fertilisation with native species, the evolution of 'super weeds', and prospective effects on human health.

GOVERNANCE

In the academic literature, some interpretations of 'governance' refer to the full array of agents – public and private – that are involved in the conduct, regulation and administration of social, economic and environmental affairs (for example, Doel, 1999; Jessop, 1995). Here I take a more circumscribed, government-centred interpretation, while acknowledging that a distinctive feature of contemporary governance is the devolution of roles and responsibilities to non-state parties; for example, to the private sector through self-regulation, to communities in the form of programs like Landcare and, through the transfer of risk, from the state to individuals (for example, crop insurance). Even a government-centred view opens up a vast array of policies, regulations, programs and strategies that relate to natural resource management, primary production and rural communities. Two spheres of government policy that have direct bearing on the relations between primary producers and natural capital are considered briefly here – policy relating to natural resource management and farm sector policy.

The governance of natural resource management operates at levels ranging from the international through to the local level. Within Australia, governments have generally progressed towards an acceptance of the extent and severity of environmental degradation. There has also been a diversification of the strategies, instruments, methods and approaches to natural resource management. In particular, associated with the emergence of neoliberal policies has been a shift towards the clearer demarcation of property rights in resources, greater use of market mechanisms for the allocation of natural resources and in support of environmental quality objectives, and a devolution of responsibility to individuals, communities and the private sector for natural resource management. Thus:

> These programs reveal an increasing federal government commitment to ecological sustainability and awareness of the need to address major environmental issues, accompanied by a trend towards transfer of responsibility for conservation to land managers and the community. (MRAP, 2001: 22)

Recent reviews of federal government strategies, such as the Natural Heritage Trust (NHT), stress community–government partnerships, regionally specific approaches, the importance of 'capacity for change' (Lockie et al., 2000a) and 'capacity building for improved natural resource management' (NNRMPS, 1999). As Higgins and Lockie (2001b: 103) point out, though, the rhetoric of empowerment is underpinned by an agenda of economic rationalism:

> NRM programs do seek to 'empower' land managers, but only in ways that are consistent with economically calculable and manipulable forms of knowledge. The effect is a rather restricted focus on environmental responsibility and sustainability as individual economic problems that can be addressed solely through improved business management.

A topical issue in relation to the governance of natural resources is that of property rights. Through time, there have been increasing expectations in terms of the *duties* and tighter curbs on the *rights* of landowners in relation to their use of land and natural resources (see, for example, Mobbs & Moore, 2002). Broadly, there are two main trends in the definition and interpretation of property rights in relation to the environment and natural resources. One is towards improved demarcation of rights to resources, including, for example, the separation of rights to land and water, which is central to the COAG[4] reforms in Australia (see, for example, Young & McColl, 2003). The assumption is that, in a framework of clearly specified, private and tradeable entitlements, natural resources will be allocated towards the highest value uses. At the same time, the rights must be appropriately priced, to ensure that they are not used inefficiently (see Young & McColl, 2003 for a detailed treatment of these issues). The second broad trend in terms of property rights is towards curtailment of rights/increase in responsibilities in respect of the use of private property. Infringements on property rights typically engender strong opposition from landholders.

A second area of governance that has implications for natural capital and its use by primary producers is farm and economic policy. A key element of economic and market reform over the last decade has been the National Competition Policy (NCP). The NCP reform agenda has been driven primarily by Australia's declining economic performance and particularly by concerns over inflation, unemployment and large current account deficits (Forsyth, 1992). The NCP has considerable implications for land and water management, indirectly through reforms affecting primary industries (see, for example, Cocklin & Dibden, 2002a, 2002b) and specifically as it applies to natural resources (for example, water reform).

Government policy in respect of risk management by producers is also relevant. Compensation payments by governments to producers for losses incurred as a consequence of extreme events have traditionally insulated farmers and other primary producers from the risk of economic loss. However, the new philosophy emphasises building risk associated with extreme events into farm plans, thereby shifting the burden of risk to individuals. The underlying philosophy is consistent with the neoliberal paradigm, which emphasises the 'responsibilisation' of individuals. While 'exceptional circumstances' assistance is still available, 'exceptional circumstances support represents a crucial means of constituting environmental risk as a business management issue that can, in most cases, be properly addressed via individual endeavour and formal planning practices' (Higgins & Lockie, 2001b: 101).

The directions in farm and economic policy can have implications for natural resource management in other ways. Recent international literature on farm-level responses to policy and regulatory changes, for example, indicates a range of possible adjustments (for example, Bradshaw et al., 1998; Chiotti, 1995; Knutson et al., 1997). This farm adjustment literature talks about diversification, which is taken to include both changes in production (for example, to other types of product, new management systems, organics, and non-traditional enterprises such as tourism), as well as the transfer of labour to non-farm activities. Land use change has implications for patterns of both land

and water use, while the reassignment of labour to non-farm activities raises issues about the allocation of effort to environmental management activities (both property and community-based). There is also debate as to whether the removal of subsidies and deregulation leads to better or worse environmental management among farmers (Bradshaw et al., 1998). Some have argued that withdrawing subsidies is good for the environment, since it discourages farmers from using artificial fertilisers and other chemicals. The counter-argument is that lower prices drive the need for higher productivity and hence increased use of water and chemical inputs; once farmers are exposed to greater price volatility, their interest in environmental management wanes – in accordance with the adage that 'it's hard to be green when you're in the red' (Cocklin et al., 2003; Martin & Woodhill, 1995).

LAND USE CHANGE AND OWNERSHIP

From the perspective of land and water management, changes in land use have potentially significant implications. Changes in land use can be brought about by a wide range of factors, including expanding or contracting markets for primary products, changes in the prices of commodities, and patterns of domestic and international investment. The expansion of plantation forestry is one example of changing land use occurring in several Australian states. Information from the National Plantation Inventory (NPI) documents a considerable expansion in the planted area, especially of hardwoods (blue gums), largely on private land (BRS, 2000), driven by improved international markets (Hopkins, 2001; Gippsland Development Ltd, 2001).

The Private Forestry Council of Victoria recently indicated a new emphasis in policy, with a shift away from explicit planting targets to the express inclusion of social and environmental issues; that is, sustainable private forestry development. Principles of the strategy are reported to include community involvement in decision-making, the sustainable management of native forests on private land and a commitment to dealing with degraded land (Hopkins, 2001). The potential for a complementary relationship between the commercial aspects of forestry plantations and the environment has been demonstrated by the oil mallee industry in Western Australia, which Tonts and Black (2003: 120) argue:

... has the potential not only to ameliorate land degradation and diversify farm incomes, but also in the longer term to reduce dependence on non-renewable energy resources and to reduce greenhouse gases by sequestering carbon.

The expansion of plantation forestry is not without its potential downsides, however. Research on plantation forestry in New Zealand and Australia has documented a plethora of perceived and real social impacts, including rural population decline, impacts on rural land values, concerns about land ownership, employment effects, and loss of income to agricultural processing industries and associated infrastructure (Barlow & Cocklin, 2003; Cocklin & Wall, 1997; Wall & Cocklin, 1996; MAF, 1994; Institute of Land and Food Resources, 2000). Additionally, New Zealand research has identified negative environmental outcomes, including effects in terms of soil erosion, rainfall interception, soil nutrient depletion, water quality, spray drift and the spread of weeds and pests (Maclaren, 1995, 1996).

Market conditions have played a major part in the expansion in plantation forestry, but government policy can also act as a key causal factor in land use decisions. The recent deregulation of dairying, for example, has the potential to bring about shifts in land use (Cocklin & Dibden, 2002b). As noted above, the international farm adjustment literature has presented evidence of various types of adjustment among primary producers to policy initiatives, including changes in inputs and outputs, and the reallocation of labour – for example, to off-farm activities (see, for example, Ilbery & Bowler, 1998).

Other drivers of land use change are relevant. In the context of analysing the comparative strengths of Gippsland for dairying, for example, it was predicted that:

> The rising cost of water will mean that northern dairy farmers won't be able to compete for water resources against higher value industries such as horticulture. Environmental problems such as salinity, rising water tables, and deteriorating water quality in northern Victoria may mean that dairying in that district is unsustainable in the future. (Gippsland Development Ltd, 2001: 16)

This pessimistic forecast has been supported by the reports from this region of financial stress resulting (anecdotally) in the sale of water rights by 10 per cent of dairy farmers during the recent drought.

The ownership of land is also potentially significant. There are at least two interesting aspects to this; a possible trend towards more leasing of land by farmers, and the prospects of more ownership of land by corporate interests. The two may not be independent of each other. In terms of the first, at the 2001 Gippsland Dairy Conference the issue of leasing rather than owning assets was a popular theme. Phillip Ruthven, Chair of IBIS Business Information Pty Ltd, opened up the discussion in his plenary paper by proposing that successful businesses would divest assets in preference for leasing (Ruthven, 2001). Leasing is not common within agriculture, but it may become more so. If that were the case, it is likely that there would be issues about long-term land management for sustainability.

The other aspect of ownership is an increasing involvement of corporates. In a SWOT analysis of dairying (Gippsland Development Ltd, 2001), it was commented that:

> There is little indication that farmers' children are interested in an agricultural/farming lifestyle. It is therefore probable that the next generation of landowners will be corporate bodies, perhaps overseas owned, *which may not have the desired commitment to sustainable land management*. (emphasis added)

Overseas analysis of the implications of industrialised agriculture in terms of sustainability principles suggests that such trends might indeed be cause for concern (Troughton, 2002; Furuseth, 1997). In respect of hog production in the United States, for example, Furuseth (1997: 308) concluded:

> The sustainable agriculture paradigm requires that capital be passed from one generation to the next. The integrated industrial farm that is the template for American hog production does not meet this test. This efficiency driven system of animal production has sharply redefined interrelationships between hog growers and their neighbours and created widespread conflict at the local, regional and state levels. Human-made capital has replaced value assigned to natural capital and superseded cultural capital.

Industrialised agriculture of the type that Furuseth refers to is not widespread within Australia as yet, and what there is tends to be centred on a few commodities (particularly cotton and rice). Nevertheless, the prospect exists that corporate ownership of agriculture and the industrialisation of production will increase.

NATURAL CAPITAL, SOCIAL CAPITAL AND COMMUNITY SUSTAINABILITY

Pastor (2001) was interested in the prospective relationships between social capital and natural capital. These differed according to the kind of social capital present, with a distinction made between 'bonding' social capital (relationships within a social group and with friends and relatives) and 'bridging' social capital (relationships with people from other social groups that may differ according to such things as ethnicity, religion or socioeconomic status). On this he commented that:

> ... the lack of bonding social capital within a community can lead to environmental vulnerability; organizing, on the other hand, can lead to direct improvements. Yet the bridging aspect of social capital may be just as important for enhancing natural assets. (Pastor, 2001: 14)

The connections between natural and social capital that Pastor (2001) refers to have similarly been identified by others. According to Berkes and Folke (1993), for example, a condition for sustainability is a supportive relationship between natural capital and society. Pretty (1999) points to the common features of both natural and social capital – both are the foundation of economic growth and human welfare, both tend to be public goods, and both can be degraded by 'externalities', the costs of which are generally borne by society at large. However, he argues that 'natural capital and social capital can be regenerated, recreated and reinforced ... Sustainable agriculture represents an important entry point for rebuilding natural and social capital' (Pretty, 1999: 7). More recently, he has suggested that:

> ... it is now clear that new thinking and practice are needed, particularly to develop and spread forms of social organisation that are structurally suited for natural resource management and protection at local level. (Pretty, 2002: 73)

In Australia, one of the distinctive ways in which the links between natural and social capital are forged in rural areas is through community involvement in environmental management and remediation programs, including the high-profile Landcare program. In many respects Landcare and similar programs have been a success, achieving useful outcomes for both communities and the environment. They encourage communication and co-operation among stakeholders, they serve as effective mechanisms for community education, and have also been used with good effect to improve the skill-base of community members (Curtis & Lockwood, 1998; MRAP, 2001). In the context of a study of the rural community of Guyra, New South Wales, Stayner (2003: 54) observed:

> In recent years, new human capital in the farm sector has increasingly been created through the activities of a range of farmer groups, which focus on particular issues such as natural resource management problems (Landcare, the Malpas Catchment Group), improving farm production, marketing and management. Social capital is necessary for the effective operation of these groups and is also created or maintained as a by-product of their activities.

Similarly, Dibden and Cocklin (2003: 187) found in Victoria that: 'Landcare has acted as a vehicle for a two-way flow of ideas and information, and a convergence of values, between farmers and conservation-minded members of the local community.'

The success of community-based programs is contingent upon a broad array of factors, however. The programs must take into account the particular interests, capacities and limitations of all participants and they must operate according to a set of locally relevant, yet regionally appropriate, goals and objectives. Furthermore, their success depends upon the provision of appropriate levels of managerial, administrative and financial assistance (MRAP, 2001; Conacher & Conacher, 2000). Additionally, while community-based programs may seek to align the stakeholders' interests, there may be circumstances where the interests of various participants are fundamentally at odds with each other. Existent power relations in communities can be recreated, leading Pretty (2002) to warn that the 'dark side' of social capital may 'encourage conformity, perpetuate adversity and inequity, and allow certain individuals to get others to act in ways that suit only themselves' (Pretty, 2002: 74).

Despite the prospective pitfalls, community participation has taken a central role in land and water management programs in Australia. Indeed, there have been widespread calls for the establishment of a broader range of mechanisms for community, stakeholder and regional representation in policy-making to increase the level of civil society's engagement in, and ownership of, land and water management programs (Boully, 2001; Dore & Woodhill, 1999; NNRMPS, 1999;

Wentworth Group of Concerned Scientists, 2003). Young (1997: 141) uses the term 'subsidiarity' to describe an appropriate assignment of responsibilities in relation to this strategy:

> Subsidiarity implies consultation and often the direct participation of the community and industry in decision making and implementation at local or regional levels. Rather than 'devolving' power, the emphasis is on making them accountable for the attainment of specified targets. The process is outcome oriented so that innovation at the local level is encouraged.

Boully (2001) suggests that the effective involvement of people and communities in NRM will require a real shift of power and authority to communities, along with the information that is needed to make decisions. At the same time, there is a need to develop capacity within communities. This, in turn, will depend on a shift from the 'what' – skills, abilities, and reason – to the 'how' – values, principles, behaviours and creativity:

> Changing emphasis from the 'what' to the 'how' is a fundamental shift in the way we have traditionally made decisions, managed and led within the natural resource management arenas. We are used to the 'power over', command and control model, rather than the collaborative inclusive 'power with' model that people are increasingly demanding. (Boully, 2001: 9)

A report submitted to the Government of New South Wales in 2003 also suggests a reform of the arrangements governing resource and environmental management, involving a transfer of both authority and resources to landowners and communities. In analysing the problems with current arrangements, the 'Wentworth Group' report observed that while there are literally thousands of farmers who are interested in better environmental outcomes, 'they lack resources and scientific advice and are disempowered by the existing bureaucratic environment' (Wentworth Group of Concerned Scientists, 2003: 3). In contrast, they suggest a model in which communities have much greater control, and access to resources:

> Fundamental to the success of such a model is simplifying the overwhelmingly complex structures that exist at present, to empower

the farming community to take control of the problem, to back them with first class science and provide them with adequate public funds to deliver on-ground solutions on the farm. (Wentworth Group of Concerned Scientists, 2003: 3)

Drawing the discussion back to the first part of this chapter, it is also relevant to consider the prospects for improving both social and environmental sustainability in the context of the influences on how natural capital is used. Negative influences include the erosion of social capital through demographic transitions and the various market forces that are undermining economic returns to production. However, more promisingly, public attitudes would seem to support a greater focus on environmental quality. At the same time, the 'tool kit' of interventions in the form of governance and regulation of the environment is developing in interesting and effective ways: the clearer specification of property rights, the better design of market mechanisms, improved information and knowledge exchange, and improved institutional arrangements offer the prospect of much better governance. And, while there are concerns about the social and environmental aspects of new technologies, such as GMOs, others (for example, more efficient irrigation systems) present the promise of improved agricultural sustainability.

There is little doubt that there is potentially a mutually reinforcing relationship between natural and social capital. In spite of the documented problems of community-based involvement, there is a widespread view that it has been effective in strengthening communities. There is also a view that communities with high levels of social capital are those best placed to effectively manage natural capital. The twofold challenge lies in: (a) discovering what is required to develop social capital, and (b) developing the strategies for bringing this capital to bear in the most effective ways in support of the management of natural resources. There would seem to be plausible strategies for both in the context of natural resource management in Australia. What remains to be seen is whether there are the political interest and the resources to ensure that the mutually reinforcing capacity is realised.

CONCLUSION

The use of natural capital by farmers is influenced by a variety of factors. As the brief overview in this chapter indicates, these factors draw the relationship between production and environment in quite different directions. Some, such as community attitudes, add impetus to the imperative to improve environmental performance. Environmental degradation itself can similarly motivate farmers to manage natural resources in a more ecologically sustainable way, particularly as the understanding by landholders of its extent and impacts improves. Other influences exacerbate the exploitation and degradation of natural resources, continuing a pattern of resource depletion that was set in place from the time of European settlement. In many cases, the signals are ambiguous. On the one hand, consumers seem to have an interest in commodities and services that are less environmentally damaging, yet there is reasonable doubt as to whether they would accept prices for food and fibre that internalised the real costs of production. Governments regulate in favour of improved environmental outcomes, but at the same time champion the 'productivist' model of agriculture, driven by ideological commitments to competition, efficiency and free trade. For farmers, it is a perplexing mixture of messages.

On balance, it is difficult to envisage a future in which the environmental performance of Australian agriculture will improve by any meaningful measure. This is not necessarily due to a lack of will among farmers, many of whom are all too aware that the depletion of natural capital ultimately threatens the long-term sustainability of their farms and, by extension, their communities. However, unless there can be a constructive progression away from the productivist model that currently dominates, it is difficult to envisage a more environmentally beneficent agriculture. Such a progression would need to be advanced on many fronts, but it would have to include strategies to build improved relationships between social and natural capital. The international literature presents a view that the two are mutually reinforcing; the pursuit of a sustainable agriculture could mean an improved natural environment and improved social and community capital. Landcare and similar community-based environmental programs seem to substantiate such claims. This is not to suggest that Landcare and other programs are not without their problems – they are inadequately resourced, there is an increasing concern over participant 'burn-out', Landcare groups have sometimes entrenched powerful groups to the exclusion of other participants, and there is doubt over the real ecological benefits, particularly when a wider landscape perspective is taken. Recent reflections on Landcare and community-based involvement generally, though, point towards prospective improvements, based in a *real* transfer of authority to the community level, and a substantial upgrade of resources (financial support, science, community capability). If these were to occur, then both natural and social capital would benefit in mutually supportive and substantive ways. This would be one important step in the direction of a more sustainable agriculture for Australia.

NOTES

1   For an overview of the early responses to the article, see *Nature*, 1998, Vol. 395, No. 6701, p. 430.

2   In the 1999 ABS survey, health was rated the most important social issue by 30 per cent of those surveyed, crime by about 25 per cent, education by approximately 16 per cent, and unemployment by 14 per cent.

3   J Sainsbury plc. At: <http://www.j-sainsbury.co.uk/index.cfm> (24 March 2004).

4   A 'Water Reform Framework' was agreed in February 1994 by the Council of Australian Governments (COAG), which consists of the Prime Minister, Premiers and Chief Ministers of states and territories, and the president of the Australian Local Government Association. It established a timetable for action aimed at halting environmental degradation and minimising unsustainable use of water resources. The Framework includes provisions for water entitlements and trading, water pricing and institutional reform. See discussion in Young and McColl (2003).

PART III

RESPONDING TO THE CHALLENGES

11

GOVERNMENT POLICY AND RURAL SUSTAINABILITY

Matthew Tonts

The changes facing Australia's rural communities are the result of complex interactions between ecological, economic, social and political forces. One of the most significant components of these is government policy. The purpose of this chapter is to examine the role of government policy in promoting a sustainable rural Australia. Since it is obviously not possible to cover all aspects of government policy within a single chapter, its aim is to provide a broad overview of how governments have approached questions of rural development and sustainability. The concept of sustainability has only recently been adopted by governments in relation to rural areas, and represents an important shift in thinking. Rather than focusing narrowly on issues such as economic development and productivity, sustainability provides an opportunity for governments to look at links between economic, social and ecological systems when formulating and implementing policy (Brunckhorst, 2002). A key element of sustainability is a consideration of intergenerational issues, particularly in relation to the availability of economic, social and ecological resources/services. Thus, a central issue facing government is not only how to tackle problems in an integrated way (taking into account the links between economic, social and ecological systems), but also in a way that ensures a degree of equity across generations. A useful framework for considering these issues is the notion of 'capitals', outlined by Cocklin and Dibden in chapter 1 (see also Black & Hughes, 2001; Cocklin & Alston, 2003). One of the underlying themes of this chapter is to consider the role that government plays in developing or degrading natural, economic, social, human and institutional capital in rural areas. Particular attention is given to recent government interventions aimed at promoting sustainability in rural Australia.

RESTRUCTURING, POLICY CHANGE AND RURAL DECLINE

As Davison points out in chapter 3, Australian Commonwealth and state governments have a long history of promoting economic and social development in rural areas (see also Powell, 1988; Bolton, 1994; Crawford & Crawford, 2003). Indeed, for much of the 20th century, governments regarded spending on services, infrastructure and agricultural support schemes as an investment in a sustainable future for rural Australia (Taylor, 1991). By the 1970s and 1980s, however, deteriorating and often volatile conditions in the world economy began to impact directly on Australian agriculture. For farming, the key problems included: a worldwide economic recession in 1973–74; the loss of traditional agricultural markets when Britain joined the European Economic Community in 1973; subsidised overproduction in Europe and North America; rapid inflation and falling commodity prices. These changes reduced the prosperity of many Australian agricultural regions.

While governments had previously attempted to protect agriculture from these economic changes, during the 1980s a new dry ideology began to develop in which market forces, deregulation and the removal of protection were seen as the best mechanisms for promoting sustainable economic growth and prosperity. In part, this seems to have been linked to the financial problems facing governments in the 1970s and 1980s. Falling economic prosperity at the national level, largely as a result of global economic forces, decreased government revenue. Despite government attempts to stimulate the economy through increased spending in the 1970s, economic growth remained relatively stagnant. Eventually, government spending (and the taxes required to fund this) was regarded as a drag on economic growth. The outcome was a shift towards a more austere set of policies designed to liberalise the national economy and improve efficiency.

For rural Australia, this shift resulted in less intervention in the agricultural economy. Price support mechanisms, import restrictions and other tax and financial concessions were gradually reduced in an effort to make Australian farmers more competitive. There were also more direct measures, such as the Commonwealth government's Rural Adjustment Scheme, which attempted to encourage smaller, inefficient farmers out of the industry to free up land for larger, more efficient farms (Stayner, 1996). According to many commentators and policy-makers, these reforms were needed to improve the longer term finan-cial sustainability of agriculture. For example, it has been argued that farmers exposed more directly to international market forces will be forced to improve their competitiveness and profitability by adopting advanced farming methods and business practices (Gow, 1996). It is somewhat difficult to assess the final outcomes of such an approach. Certainly, many larger farms have survived and prospered, and it is these more 'efficient' farmers that governments believed would provide a financially prosperous future for agriculture. However, many profitable smaller and medium-sized producers were swept up in the calls to 'get big or get out', borrowed large sums of money to upgrade or expand their operations and were caught up in the cycle of debt and high inter-est rates of the 1980s and early 1990s (Smailes, 1996; see also Lawrence, in chapter 6).

Many of those farmers who were not financially viable left agricul-ture and often left farming districts for coastal and/or urban locations (McKenzie, 1994). This out-migration led to a reduction in spending in rural communities, a contraction of local economies, fewer employ-ment opportunities, and further depopulation. This has not only had direct impacts on the economic capital of rural communities, but also on social and human capital. The loss of population can fragment long-standing social networks and undermine the viability of organisations, such as sporting clubs, community groups and local churches. Similarly, population decline can reduce local stocks of human capital, which are generally regarded as vital for maintaining resilient and innovative rural communities (Black & Kenyon, 2001). As populations dwindle, skills, expertise and leadership capacity are often reduced, making it more difficult for communities to revitalise or even sustain themselves (Sorensen & Epps, 1996).

A range of other government policy changes affected the sustain-ability of rural communities during the 1980s and 1990s. In the drive to improve the efficiency of the national economy, many policy-makers held that revenues collected by the state for the provision of public goods services necessarily detract from private investment, saving and consumption, and thus interfere with the mechanics of the free market. The response of both Commonwealth and state governments was to reduce taxes and charges for industry and citizens. However, the capac-ity to deliver tax cuts depends, at least partly, on a reduction in the level of public expenditure. This has resulted in pressure to 'properly' cost rural services and eliminate, or at least reduce, what are sometimes seen as inappropriate cross-subsidies. Much of rural Australia has experienced the loss or rationalisation of key services, such as hospitals, schools, police stations and welfare services. This marks a distinct contrast from those earlier government policies that emphasised social equity and stability in rural communities. To many policy-makers, however, the rationalisation of services in declining rural areas is economically 'efficient', with the less desirable outcomes simply being a consequence of wider social and economic 'progress' (Lawrence, 1995). An increasing body of research has documented the impacts of this policy shift on rural communities (see Rolley & Humphreys, 1993; Cheers, 1998; Tonts, 2000; Gray & Lawrence, 2001a; Cocklin & Alston, 2003). In the Western Australian town of Narrogin, for example, the decision to rationalise the state's government railway reduced the railway workforce from over 300 in the late 1960s to less than ten in 2001. Not only did this impact negatively on the local economy, but also undermined a number of local social organisations and institutions and contributed to a substantial reduction in stocks of human capital (Tonts & Black, 2003).

The challenges for rural communities in achieving economic and social sustainability over the past three decades have been accompanied by the widespread degradation of natural capital. The clearing of native vegetation has not only reduced biodiversity and natural habitats, but has contributed to problems such as secondary dryland salinity. In Western Australia alone, it is now estimated that nearly two million hectares of land are affected by dryland salinity, with a further two million hectares under threat (Conacher & Conacher, 2000: 57). This has the potential not only to reduce the area available for agricultural production (thereby reducing productivity), but also to further degrade areas of remnant vegetation and river and stream systems. In addition, salinity is now damaging the built environment, including roads, buildings and parks and gardens in country towns (Beresford et al., 2001). Numerous other ecological problems affecting Australia's agricultural regions are outlined by Cocklin in chapter 10.

In part, these ecological impacts are a legacy of earlier government policies that encouraged (and even forced) farmers to clear natural vegetation from their properties under terms of conditional purchase (see Bolton, 1981). In addition, the focus of many government agencies on the narrow goal of agricultural productivity meant that some of the 'system-wide' impacts of farming were often poorly understood or ignored. It is clear that economic pressures on farmers have been a direct contributor to some of the ecological problems affecting rural Australia (Lockie, 2001b; Lawrence, in chapter 6). In order to remain profitable, farmers are often forced to use large quantities of fertilisers and pesticides, and in some cases to overstock their paddocks or clear more native vegetation, despite an acute awareness that land degradation is a serious problem. Indeed, research by Reeve and Black (1993) indicates that 90 per cent of Australian producers acknowledge that agricultural land is degraded. During the 1980s, land degradation became such a serious problem that state and Commonwealth governments were no longer able to ignore it.

GOVERNMENTAL INTERPRETATIONS OF SUSTAINABILITY

During the 1980s and 1990s, Australian governments began to adopt the rhetoric of sustainability, particularly following the release of the United Nation's World Commission on Environment and Development report in 1987, *Our Common Future*, otherwise known as the Brundtland Report (see Black, in chapter 2). This report had a particularly important legitimising role within government. With increasing global attention being given to environmental issues, governments could no longer continue to dismiss the problems facing natural environments. Furthermore, by emphasising the connections between ecosystems, economies and societies, the Brundtland Report forced governments to begin to think about the issues in a more integrated and systematic way. One of the first major policy responses to the Brundtland Report was the Commonwealth government's National Strategy for Ecologically Sustainable Development (ESD). Ecologically sustainable development was defined as:

> … using, conserving and enhancing the community's resources so that ecological processes, on which life depends, are maintained, and the total quality of life, now and in the future, can be maintained. (Commonwealth of Australia, 1992: 6)

While this definition emphasises the ecological dimensions of sustainability, the Strategy's core objectives provide an insight into its breadth. The three core objectives are:

- To enhance individual and community wellbeing and welfare by following a path of economic development that safeguards the welfare of future generations.
- To provide for equity within and between generations.
- To protect biological diversity and maintain essential ecological processes and life-support systems. (Commonwealth of Australia, 1992: 8)

The National ESD Strategy was complemented by the signing of an *Intergovernmental Agreement on the Environment* (IGAE) in 1992. This agreement secures Commonwealth, state, territory and local government commitments to the broad principles of sustainability. While both the ESD Strategy and IGAE were important in stimulating debate about sustainability within and among governments, their impact in the short term was questionable. Nevertheless, the documents did highlight some of Australia's most pressing ecological problems. This focus on biophysical problems was, however, often at the expense of the economic and social dimensions of sustainability. When the concept of sustainability was more broadly applied, it tended to focus on the issue of economic sustainability. Given the neoliberal policy climate of the 1990s, it is not surprising that governments emphasised the importance of maintaining growth, maximising private sector profits, and reducing unemployment (Gerritson, 2000). In this context, the natural environment was largely viewed as an economic resource that needed to be sustained. Indeed, some of the debate about sustainability was (and still is) concerned about the impact that the pursuit of ecological sustainability will have on economic prosperity (Venning & Higgins, 2001). Gradually, however, an increasing number of politicians and bureaucrats have recognised that a degraded biophysical environment has significant implications for economic wellbeing and prosperity, particularly in those areas dependent on agricultural production.

In much of the early policy discourse on sustainability, perhaps the most neglected dimension was social sustainability. While social sustainability is a rather nebulous concept, it might generally be regarded as the capacity of a society to provide for the wellbeing of its people in an equitable manner (Troughton, 1995; Yencken & Wilkinson, 2000; Dresner, 2002). Key elements of social sustainability include health, education, maintaining Indigenous cultures and social justice. For many governments in the 1990s, the notion of social sustainability tended to conflict with the prevailing neoliberal ideology. During the late 1990s, however, a shift in the mood of the electorate in some parts of the country saw hardline neoliberal policies challenged in favour of policies that recognised social needs. Much of this backlash occurred in rural Australia, where more than a decade of volatile commodity prices, out-migration, service and infrastructure rationalisation and job losses had taken their toll (Pritchard & McManus, 2000). Across the country, voters turned against governments that had neglected citizens' social needs.

At the very least, the electoral backlash in rural Australia contributed to a change in the rhetoric of government policy. A Commonwealth Regional Summit held in 1999 brought together government, business and community representatives from across the country to discuss rural issues. Not surprisingly, social equity, particularly in relation to services and infrastructure, was an important theme at the summit. At the same time, the social dimensions of sustainability were highlighted by the Commonwealth government's State of the Environment (SoE) reporting, which formed part of its obligations under the *Agenda 21* agreement (see Black, in chapter 2). During the second half of the 1990s, the UN Commission placed increasing emphasis on the social dimensions of sustainability. This meant that Australian government agencies began to report on a range of social issues as part of the SoE process. These included women, children and youth in sustainable development, strengthening the role of Indigenous peoples, and infrastructure and the built environment. Undoubtedly, this reporting helped to raise the profile of social issues in the sustainability debate, particularly with regard to government policy.

A more holistic interpretation of sustainability also began to emerge at the state government level during the late 1990s and early 2000s. In Western Australia, for example, the state government introduced a regional development strategy in 2000 that had sustainability as a core objective (Government of Western Australia, 2000: 8). The release of a revised regional development policy in 2003 continues this emphasis. These regional policies were accompanied by the development of the *State Sustainability Strategy*, which defines sustainability as 'meeting the needs of future generations through simultaneous environmental, social and economic improvement' (Government of Western Australia, 2002: 24). The document also stresses that sustainability is an integrative concept, and that an activity that meets only two of the factors simultaneously (such as economic and social but 'trades off' the environment) is ultimately not sustainable. Importantly, the policy was developed within the most powerful office in the government, the Department of

Premier and Cabinet, and is supported by a newly established Sustainability Policy Unit, which monitors government activities to ensure that sustainability principles are upheld.

While governments have certainly been keen to adopt the rhetoric of sustainability at a broad policy level, maintaining viable natural, social and economic systems depends upon positive actions by government agencies. In this respect, there have been a range of important policy developments over the past decade or so. Specific strategies have been implemented to deal with problems such as land degradation, social equity issues and economic decline. In many ways, these are about preserving, enhancing or creating various forms of capital.

RURAL SUSTAINABILITY AND COMMONWEALTH AND STATE GOVERNMENT POLICY

SUSTAINING NATURAL CAPITAL

Over the past decade, dozens of policies have been implemented at both Commonwealth and state levels to deal with issues such as salinity, soil erosion, the loss of biodiversity, and the contamination of waterways in rural (and urban) areas. One of the main problems with government approaches to these problems is that they have tended to focus on the biophysical issues, rather than considering ecological degradation as part of a wider social and economic system. In Western Australia, for example, where the problem of dryland salinity is threatening up to four million hectares of farmland, the state government's *Salinity Action Plan* (1996) focused almost exclusively on the question of ecological sustainability. It considered the hydrological and biological dimensions of the problem, and examined a range of solutions to salinity, including tree planting, changes in pasture and crop types and engineering solutions. Virtually no attention was given to how social systems intersect with biophysical systems, despite the vast body of evidence that suggests ecological rehabilitation in rural areas is dependent on a thorough understanding of local social, cultural and economic systems (Lockie & Bourke, 2001; Carr, 2002; Vanclay & Lawrence, 1995).

One of the other criticisms of the *Salinity Action Plan* is that it tended to adopt a one-dimensional view of the ecological problems facing rural areas. By focusing considerable funding and public attention on salinity, there is a risk that other equally serious problems, such as soil acidity or the loss of biodiversity, will be overlooked. This is in

contrast to the key principle within the sustainability and ecology literature that the various components of an ecosystem are inextricably
linked and that problems need to be considered on a 'system-wide' basis
(Dresner, 2002).

In tackling environmental problems on a system-wide basis,
arguably one of the success stories of government policy has been the
promotion of the Landcare movement. Landcare had its origins in
Victoria in 1986 and was adopted as a national program in 1989. With
guidance from government agencies, farmers formed community
groups, usually on a catchment basis, to deal with local and regional
land degradation problems, particularly salinity and soil erosion. The
movement grew rapidly, with more than 4000 groups operating by the
late 1990s (Conacher & Conacher, 2000). Governments have actively
supported the movement with Commonwealth funding through the
National Landcare Program and the Natural Heritage Trust. At the
state level, funding and technical support is also provided to groups
through a range of agencies and programs.

For some commentators, Landcare represents a model in sustainability thinking. It brings together communities and government in a
partnership aimed at resolving serious ecological problems. The movement not only involves government agencies, but thousands of individual landholders and community volunteers, including metropolitan
residents, with schools and local conservation groups in urban areas
engaging in various rural Landcare projects. Landcare also emphasises
the principle of intergenerational equity, by aiming to restore landscapes for future generations, both as an economic resource and in
order to provide ecosystem services and a more attractive environment.
Conacher and Conacher (2000) point to the movement's direct 'on
the ground' successes, including significant fencing and revegetation
works, soil and biological surveys, monitoring programs, and educational activities.

While the Landcare movement does have a number of shortcomings
(see Cocklin, in chapter 10; ANAO, 1997; Curtis et al., 1998), it has
played an important role in focusing attention on the need for integrated
approaches involving governments and communities when dealing
with ecological problems. A discussion paper produced by the
Commonwealth government in 1999 emphasised the benefits of this
approach. The paper, *Managing Natural Resources in Rural Australia
for a Sustainable Future*, stresses the importance of partnerships between
communities, business and government. The document also points to

the need to empower communities to deal with sustainability issues, to build on the successes of the Landcare movement, and to devolve power and authority to regional and local institutions and organisations. These suggestions were later taken up in the Commonwealth government's *National Action Plan for Salinity and Water Quality in Australia* (Council of Australian Governments, 2002).

Despite the focus of governments on preserving natural capital, there is some evidence to suggest that governments share responsibility for the ongoing erosion of this resource. In a number of states, for example, governments continue to allow farmers to clear their land of native vegetation for agriculture (for a review of current regulations, see Productivity Commission, 2003b). Furthermore, most government policies do little or nothing to alter the structural factors contributing to ecological destruction. As Lawrence argues in chapter 6, the ongoing economic pressures in agriculture, exacerbated by government policies that have exposed the industry to volatile international markets, continue to force some farmers to overexploit their land (see also Lockie, 2001b). In order to remain profitable, farmers often have little choice but to overstock, overcrop and even remove native vegetation. Furthermore, poor economic conditions have reduced farmers' ability to engage in rehabilitation projects. Virtually all of the current natural resource management strategies tend to overlook these constraints and, in devolving much of the responsibility for land rehabilitation to the local and individual level, are likely to perpetuate the problem.

THE ROLE OF SOCIAL CAPITAL

Over the past ten years or so, governments have increasingly focused on the social resources that already exist within communities that might enable them to respond to problems. In this regard, a concept that has gained considerable currency within government agencies is social capital. One of the problems with this concept, however, is definition. In many respects, the term is plagued by some of the same problems as 'sustainability', in that there are numerous competing conceptualisations and definitions. In terms of government policy, this has meant that different agencies and tiers of government have tended to interpret social capital quite differently. Most definitions of social capital tend to focus on those attributes that bind people into community and social relations that can then form the basis for action. In a recent report for the Commonwealth Department of Family and Community Services, Black and Hughes (2001) suggest that these include qualities such as trust, altruism, reciprocity, shared beliefs and norms, tolerance, a sense of belonging to a community, self-reliance and self-help (see also Uphoff, 2000).

That government agencies are attracted to the concept of social capital is hardly surprising. It focuses directly on those attributes that many communities fear are being lost in the face of rapid structural change. These same qualities also tend to appeal directly to neoliberal sensibilities by stressing the virtues of self-help, trust, sense of community, altruism and reciprocity (Badcock, 1997). The emphasis is generally on the role of individuals and communities, rather than on direct government intervention. The message to rural communities is that their future wellbeing rests largely in their own hands.

In emphasising the importance of social capital, a number of commentators have pointed to the experience of Landcare, where social bonds, trust and networks are essential for farmers to work together in a cohesive and effective manner. Indeed, a recent study of two Landcare networks by Sobels et al. (2001) suggested that the success of these organisations was based on high levels of social capital, particularly trust and a collective willingness to respond to issues. They point out that Landcare groups can also contribute to social capital by providing a network that facilitates community learning and the strengthening of local social bonds.

More recently, social capital has been incorporated into policies and programs that aim to improve the economic sustainability of rural communities. In Western Australia, for example, the state government's Regional Development Strategy suggested that economic development was directly contingent upon the presence of local social capital. In order to promote social capital, the state government provided funding for community economic and social planning workshops. The policy also stated the need to promote reconciliation and to provide support for community groups and volunteer work. It is important to note that this emphasis on self-help and community action is largely underpinned by neoliberal ideology. As in other parts of the country, the Western Australian government is committed to the principles of fiscal restraint and minimal direct intervention in the economy. Emphasising the importance of local people and resources is a key part of this strategy. There is also some evidence to suggest that this approach can be very successful. Throughout Australia, there are examples of rural communities that have been able to reverse economic and population decline, and improve the quality of services and infrastructure largely through local action and resources (Dibden & Cheshire, in chapter 12; Cocklin & Alston, 2003; Rees & Fischer, 2002; Black & Kenyon, 2001; Staples & Millmas, 1998).

Despite the increasing focus on social capital by governments, there is considerable evidence to suggest that many current and recent policy decisions are actually eroding rather than enhancing social capital. For example, policies that contribute to the withdrawal or ratio-nalisation of services can remove individuals and families from rural communities and destroy social networks (Gray & Lawrence, 2001a; Gerritson, 2000; Haslam-McKenzie, 1999). Similarly, uncertainty about the future of government services can erode trust and confi-dence, thereby depreciating stocks of social capital in rural communi-ties (Haslam-McKenzie, 2003c). So, while it is convenient for governments to talk about the importance of building social capital as a way of promoting strong and vibrant rural communities, it is also important that they are aware of the negative impact that they can (and often do) have on this resource.

HUMAN CAPITAL

The concept of social capital is often accompanied by discussions about human capital. This largely refers to the status of individuals, and to the health, education, skills and knowledge of people (Black & Hughes, 2001). It also takes into account qualities such as leadership, entrepreneurialism and innovation. One of the elements of human capital that has received significant attention is leadership. This has been a particularly important component of the Commonwealth government's Stronger Families and Communities Strategy. This program has been allocated funding of $37 million over four years to identify and support potential community leaders in socially disadvan-taged areas across Australia. Other programs include the Australian Indigenous Leadership Program, which aims to assist Indigenous people who are active in their community's affairs to participate in leadership training schemes. A number of programs also exist at the Commonwealth level that aim to promote leadership among young people. For example, the Department of Agriculture, Fisheries and Forestry has funded the Young People's Industry Leadership Project, which aims to increase the capacity, involvement, experience and profile of young people (18–35 years) in rural industries. This program involves a number of key elements, including an industry leadership course and a rural youth conference.

Similar leadership programs have been developed at the state government level. In Western Australia, the State Regional Development Policy argues that:

> ... skilled motivated leaders can encourage involvement and the development of projects and networks ... The development of leadership enables communities to ... capture development opportunities that sustain viable settlements and business enterprises. (Government of Western Australia, 2000: 13)

In order to foster leadership, the state government initiated a range of training programs that focused on regional entrepreneurialism, decision-making, structures of governance and citizenship.

While there is evidence that leadership can be an important element in promoting more sustainable rural communities (for example, Sorensen & Epps, 1996), there are a number of clear limitations. Firstly, many of the programs tend to focus on individuals who already demonstrate some leadership capacity, and rarely focus on 'latent' leaders. Often, those who participate in leadership programs are people from high-income and status groups, leaving serious question marks over the inclusivity of the programs (Lannin, 1997). Secondly, the small size of some rural communities means that the chances of finding a pool of potential or existing leaders can be problematic (Sorensen & Epps, 1996). Finally, while effective leadership can help to stimulate successful local projects, these are often affected by wider structural and policy changes that can undermine their effectiveness (Haslam-McKenzie, 2003c).

The focus on leadership by governments is just one aspect of human capital. Education, knowledge, skills and the health status of residents are all components of human capital and are critical in maintaining sustainable communities. In the case of Narrogin in Western Australia, for example, local skills and expertise were critical in developing a successful revitalisation strategy in the late 1980s and early 1990s (Tonts & Black, 2003). In many communities, however, government policies are actually undermining human capital, since the withdrawal and/or rationalisation of services tend to reduce stocks of this resource as public sector employees leave town. Indeed, the rationalisation of services, such as health and education, has the capacity to remove considerable stocks of human capital, as these sectors often consist of some of the most educated components of the labour force. The associated negative multipliers that public service withdrawal generates can further erode human capital if businesses close and the owners of these firms leave the community.

INSTITUTIONAL CAPITAL

Traditionally, governments played a critical role in building up institutional capital in rural communities (Tonts & Black, 2003). Government institutions (and their employees) usually spend considerable sums of money in the local and regional economies, providing an important source of income for small businesses. Indeed, in times of economic stress in sectors such as agriculture, government institutions often provide country towns with a valuable economic backstop. As Hudson (1989) has demonstrated in rural New South Wales, the spending of public sector institutions and employees is a critical component of local economies and can, to some extent, soften the impacts of poor seasons in agriculture.

In addition to economic benefits, public sector institutions contribute to the sustainability of rural communities in a number of other ways. Indeed, it might be argued that some of the most critical institutions in sustaining rural communities are those that:

- ensure the physical and mental health and wellbeing of individuals, such as hospitals, aged care facilities, preventative health care programs and counselling services
- build human capital through education and training, such as schools, TAFE, distance learning programs and various government training programs
- preserve or enhance natural capital, such as conservation agencies

However, it is these institutions that are often most at risk of government rationalisation or withdrawal. This not only has immediate economic impacts, but a range of broader social, cultural and environmental implications, both in the short and long term. As shown above, the loss of the people employed in these institutions can undermine stocks of social and human capital in rural communities. Furthermore, there is evidence to suggest that a number of the serious health, education and environmental issues facing rural communities are the direct result of the absence or withdrawal of government institutions (HREOC, 2000b; Gerritson, 2000; Rolley & Humphreys, 1993).

While state and Commonwealth government institutions are critically important to rural sustainability, it is often local government that is at the forefront of efforts to maintain the economic, social and ecological viability of rural areas. Traditionally, Australian local governments have been responsible for local planning, environmental health, basic services and infrastructure. Secondary roads, rubbish collection, libraries, sporting facilities and community meeting halls are just some of the services and infrastructure that local government provides. These are usually funded by local rates and charges, and grants from state and Commonwealth governments. Over the past two decades, however, the responsibilities of local governments have steadily grown.

This changing role for local governments is, in part, linked to shifts in Commonwealth and state government policy. In a number of areas, these higher tiers of government have attempted to reduce spending by devolving certain aspects of service and infrastructure delivery to the local level. Local governments have found themselves responsible for some health care services (especially for the aged), housing development, industrial infrastructure and telecommunications services. While the rhetoric of state and Commonwealth governments is that local governments are usually more attuned to the needs of local people, the reality is that by devolving these responsibilities governments are able to cut costs in line with the principles of economic rationalism (Tonts & Jones, 1997). This helps to explain the growing financial pressures facing local governments across Australia (see Daly, 2000). The risk for residents is that their local government may not provide some services or facilities available elsewhere, which has the potential to contribute to inequalities between regions and localities.

Another area in which local governments have taken an increasing interest is local and regional economic development. There are two interrelated reasons for this. Firstly, Commonwealth and state governments have argued that local governments are better able to identify local needs and priorities, can better utilise local skills, and can increase participation in the planning process (Beer et al., 2003). Secondly, the absence of effective and substantial state and Commonwealth action in dealing with the economic and social problems facing rural areas has meant that local governments have been forced to deal with problems themselves. This mainly occurs because local government comprises resident councillors who are directly affected by the economic and social changes impacting on rural communities, such as population decline, economic contraction and service withdrawal.

It is, however, important to recognise the limitations of local government as a contributor to institutional capital. The actions of local government are often dependent on the stocks of locally available human, social and economic capital. If some or all of these forms of capital are weak, then there is a risk that the effectiveness of local governments will be reduced. For example, an absence of social capital as a result of a lack of trust in community leadership has the potential to undermine local development initiatives, because small business owners and residents may not support the actions of the council. Alternatively, low levels of economic capital may limit the financial resources that can be raised for the provision of local services. While the notion of 'capitals' has rarely been used to analyse local government, there is a clear body of evidence to suggest that, without attributes such as leadership, community backing and adequate human and financial resources, local governments will struggle to reverse major structural economic and social problems (Sorensen & Epps, 1996; Forth, 2000; Beer et al., 2003). Indeed, even with these resources, the chances of local governments succeeding on their own can be remote.

GOVERNMENT AND THE INTERCONNECTEDNESS OF CAPITALS

One of the ongoing problems with government policy is that it often fails to recognise the interconnections between various forms of capital. This is despite the considerable evidence demonstrating that ensuring sustainability of rural communities requires a holistic and integrated approach. Indeed, one of the most valuable aspects of the concept of sustainability is that it stresses interdependence and interconnections (Dresner, 2002). It recognises the close links between natural, social, human, institutional and economic forms of capital. However, there is still a tendency within government to isolate issues and areas of responsibility within narrow portfolio areas. In some ways, this is not particularly surprising. It is administratively and financially convenient, and continues a long legacy of 'reductionist' policy-making. So, while concepts such as social and human capital are now widely recognised by agencies responsible for portfolio areas such as regional development, community services and health, they often have little currency within those organisations responsible for natural resource management. Similarly, few programs that deal with issues such as rural health have attempted to understand the links between natural capital and the individual wellbeing, overlooking or despite evidence that suggests close connections between natural capital and human health (Rapport et al.,

1998). A greater recognition of the complex links between the various elements of rural systems seems likely to produce more effective policy responses (Brunckhorst, 2002).

There are, however, some signs that more holistic approaches to rural policy and planning have begun to emerge. In Western Australia, for example, the Sustainable Rural Systems Program – launched by the Department of Agriculture in the mid-1990s – was designed to bring together government agencies to develop integrated approaches to ecological and social problems in rural areas. There was an explicit recognition that environmental issues could not be solved or tackled without dealing with problems such as population decline, financial hardship, service withdrawal and infrastructure deterioration. In attempting to deal with these issues in an integrated way, the program focused on improving leadership and education, building social networks, enhancing employment opportunities, promoting economic development, and sustainable land-use planning. For a department previously focused almost solely on farm production, this was a radical shift in direction. The outputs of the program included documents and policy statements relating explicitly to natural, social, human, economic and institutional capital and the close interconnections between these.

This integrated approach was not without problems. By adopting such a broad approach to understanding and managing rural systems, other agencies often felt that their 'turf' was being invaded, and that it was beyond the responsibilities of the Department of Agriculture to become involved with issues such as regional development, planning and community services. Following a change of state government in 2001, the Sustainable Rural Systems Program was dismantled and the Department of Agriculture encouraged to focus on its 'core business' of farm productivity. The outcome has been a return to narrow, sectoral approaches to rural policy. While other agencies across Australia (for example, the Murray-Darling Basin Commission) are grappling with ways of developing integrated approaches to rural sustainability, the portfolio-based approach continues to dominate, in spite of its obvious shortcomings.

CONCLUSION

Despite changing ideologies and approaches to public policy over the past three decades, governments continue to play a crucial role in the pursuit of a sustainable future for rural Australia. The renewed interest in issues associated with rural service provision, infrastructure and community wellbeing has brought into focus some of the problems associated with policies focused on economic efficiency and fiscal restraint. While there have been few indications that governments are about to embark on major spending programs in rural areas, there does seem to be a recognition among an increasing number of bureaucrats and politicians that current policies are often undermining the social and economic sustainability of rural areas. In response, a number of lightly funded 'remedial' strategies have been implemented to deal with some of the problems facing rural communities. As part of these, the importance of maintaining social, human, institutional and economic capital is slowly being accepted as an essential component in ensuring the sustainability of rural communities.

Similarly, there has been a recognition that current patterns of development are ecologically unsustainable. Governments at all levels have committed significant financial resources to tackling problems such as land degradation and loss of biodiversity. At the heart of most of these policies are explicit statements about sustainability. More often than not, however, these statements are focused on the biophysical elements of sustainability. Typically, little recognition is given to the close links between economic, social and ecological systems, despite the overwhelming evidence that actions to support ecological sustainability demand an understanding of the importance of economic and social sustainability. In many cases, government policy is even quite contradictory, with some policies degrading various forms of capital, while others aim to promote their rehabilitation. A case in point is the erosion of natural capital as a result of government policies that allow land clearing. This is occurring at a time when other policies are aiming to promote biodiversity and landscape rehabilitation. What needs to be appreciated is that focusing on one aspect of sustainability while undermining others is, ultimately, unsustainable. Thus, one of the key challenges for governments in the 21st century is to begin to address issues of sustainability in a genuinely integrated way. Given the current segmented approach that currently dominates government policy, there is a considerable way to go until meaningful policy integration and coordination is achieved.

12

COMMUNITY DEVELOPMENT

Jacqui Dibden and Lynda Cheshire

The problems facing Australia's rural communities are, by now, well known. Since the 1960s, when it first became evident that the rural sector was under strain (Linn, 1999), Australian governments have searched for a solution to the rural downturn in the hope of making rural Australia economically, socially and environmentally viable. However, with neoliberalism as a guiding policy framework from the mid-1980s, many strategies have tended to prioritise economic growth at the expense of approaches that provide rural citizens with equitable access to social and welfare services. This has done little to promote community wellbeing on a broader scale or to overcome the social exclusion experienced by many rural dwellers.

Over the last decade, therefore, a renewed focus has been placed upon the interrelatedness of the economic, social and environmental dimensions of rural decline and, accordingly, on strategies of rural community development to deal with these issues in an integrated manner. These new approaches clearly embrace the language of the social by making 'community', as opposed to simply economic, development a prime objective. Precisely what this means, however, remains unclear, particularly since this new form of community development bears little resemblance to the social change mandate of the earlier community development movement. While both approaches share a common discourse of 'community', 'self-help', 'bottom-up' and 'empowerment', the assumptions and objectives underpinning them are vastly different. The purpose of this chapter, therefore, is to examine precisely what is meant by the term 'community development' and the various ways it has been understood and applied by policy-makers and practitioners. Since 'community' and 'development' are both much-contested terms (Robbins, 1981), the task is not straightforward. Nevertheless, it remains important because it brings to light the differing, and often competing, rationalities of economic growth and social inclusion that coexist within the rural community development approach. In doing this, the chapter undertakes a critical analysis of contemporary rural community development initiatives in Australia to examine the policies, practices and implications of this approach.

INTERPRETATIONS OF COMMUNITY DEVELOPMENT

The concept of community development has been interpreted in a number of ways according to the philosophical traditions of its various proponents. The uncritical usage of the term has caused some confusion between and, indeed, conflation of quite distinct approaches of economic and social regeneration. To clear up this confusion, Beer and Maude (2002) point to two ways in which community development has been understood in Australia. In the first, community development is treated as a form of economic development whereby communities seek to improve their own wealth and employment potential through the development of local resources. The inclusion of community into what is essentially a process of economic development makes reference to both the scale at which that development takes place – most often a geographic area such as a neighbourhood or locality ('a community') – and to those who should become involved in, or initiate, that development ('the community') (Gibson & Cameron, 2001; see also Keane, 1990). More accurately described as community economic development or local economic development, the emergence of this form of development in Australia may be seen as a response to criticisms that previous, state-sponsored approaches were excessively preoccupied with attracting external capital, but did little to include local communities in the decision-making process (Gibson & Cameron, 2001; Garlick, 1997). While the emphasis has shifted to supposedly bottom-up processes of community participation in development activities, the proposed solutions generally remain economically driven and based upon the assumption that environmental and social benefits will 'trickle down' from job creation and local industry development.

The second definition of community development has a similar emphasis upon collective action but, in coming from a social work or community activist tradition, seeks to tackle social alienation and exclusion more explicitly. Advocates of this approach – which arose in Australia in the 1970s – point out that market-based remedies to community decline are part of the problem, not the solution, because they emphasise efficiency over equity, and competition over social justice (Ife, 1995). In its most liberal form, community development attempts to address the economic and social fallout of the market model by providing social and welfare services to those who are most disadvantaged by, or excluded from, mainstream society (Cheers & Luloff, 2001). Where once these ameliorative strategies were targeted primarily at individuals and families, the recognition that whole groups or localities can experience disadvantage has strengthened the case for community development to enhance the quality of life of the entire community. The problem with this approach, however, is that it risks ignoring the social exclusion of particularly disadvantaged residents within that community.

From a more radical perspective, community development is not simply about improving the lives of community members through top-up measures of welfare (Cheers, 1998) – it is a form of political action (Warburton, 1998) that transforms people's lives by targeting the structural causes of inequality and disadvantage. Thus, community developers seek to redistribute power and resources to disadvantaged groups (Haughton, 1999), to foster collective action through processes of consciousness raising (Burkey, 1993) and, ultimately, to empower marginalised populations to challenge dominant ideas, structures and institutions (Herbert-Cheshire, 2000). For proponents of this approach (Ife, 1995; Labonte, 1999; Cheers & Luloff, 2001), community development is an end in itself, enhancing the skills and confidence of the local community to articulate its own concerns and visions.

In recent decades, however, the idea of building community to cure the ills of society has become increasingly popular in governmental discourses, as evidenced by the surge of community-based solutions to problems as diverse as environmental decline, crime control and aged care. The incorporation of community development into the hegemonic discourse of the state has led to the dilution of its radical force and an increased focus on its instrumental, rather than empowering, potential. In other words, community development is no longer seen as an end in itself, but rather as a useful means of mobilising people to achieve various other goals – such as the withdrawal of government-provided services through the promotion of community self-help schemes (Murdoch, 1997), or the legitimation of policy decisions through claims that 'the community' has been consulted (Smith, 1998; Lawless, 2001). When applied in this way, community development is no longer a political activity; indeed, it may be considered apolitical since it redirects attention away from a critique of government decisions towards the provision of local services. This has caused some authors to express concern about the co-option of community development into the state's own, neoliberal, agenda as a 'sugar coating for a bitter pill' (Bryson & Mowbray, 1981: 263).

In spite of these different interpretations of community development in Australia, some common themes distinguish community development from a straightforward economic approach. The first of these is the seemingly obvious assumption that there is something called a 'community' to develop, or at least to participate in the development process. Moreover, in contemporary discourses of development, the plight of declining areas is frequently understood in terms of a 'loss of community' (Rose, 1996; Herbert-Cheshire, 2000), to which solutions of community-building or community development must be applied. Debates over the precise nature and definition of community are long-standing in the academic arena (see Black, in chapter 2), with most authors agreeing that 'physical proximity does not always lead to the establishment of social relations' (Stacey, 1969: 144). Nevertheless, the recent emphasis upon community in development discourses is one that not only associates community with locality (Gibson & Cameron, 2001), but also expects that feelings of cohesion can be found, or fostered, among the people who live there. The idea that community is as likely to be a site of conflict and exclusion is not always considered.

The second implicit assumption is the need to involve the community in its own development through strategies of community self-help. While this may amount to little more than consultation, it is expected that community development will be 'bottom-up' rather than 'top-down' and 'endogenous' rather than 'exogenous' (Friedmann, 1992), drawing upon the skills and resources that already exist within a given community, rather than seeking to attract external investors. Such an approach is said to be more sustainable in the long term because it creates a process of development that is embedded in local social and institutional patterns (Day, 1998b), and relevant to local needs and conditions (Maude, 2002).

A third point of agreement is that, where skills and resources are found to be lacking, and community 'capacity' or 'capability' for self-help is low, programs to 'build' community capacity are required (Ashby & Midmore, 1996: 114). A variation on this approach argues that the capacity required is not the ability to develop internal resources, where these do not exist, but being empowered to access resources from outside the community on grounds of equity. The concept of 'capacity building' as a key feature of contemporary strategies of rural development in Australia is explored later in this chapter, yet it is worth pointing out here that, like community development, the term is subject to various interpretations and has been seen as serving both instrumental and empowering ends.

The need for capacity building in communities of decline or disadvantage raises a fourth issue concerning the role of the state and other outside parties in community development activities. Among community development proponents, there is broad consensus that some level of state support is required, although the nature and extent of that support is often contested. For writers such as Haughton (1999), who emphasise the structural causes of decline and disadvantage, community-based approaches to development are important but, on their own, cannot deal with the broader social, political and economic structures over which communities have no control. For this reason, Haughton believes that community development should complement, rather than replace, state expenditure in the development process. Others, who support a neoliberal policy agenda, see state intervention as exacerbating problems because it creates welfare dependency among the population (see Ife, 1995, for a critique of this argument). As a result, they promote a reduced, but nevertheless important, role for governments in providing the political and institutional frameworks within which self-help and self-reliance can flourish. In practical terms, this involves a shift away from the direct provision of state subsidies to declining areas towards an investment in the skills, resources and information that enable people to pursue their own development goals (Western Australia Department of Commerce and Trade, 1999).

RURAL COMMUNITY DEVELOPMENT

While economic and social development are increasingly intertwined, it is still possible to identify a clear separation of the two in Australian government policy. Indeed, a lack of integration has been noted between the two streams of economic and social development as they are pursued by different government agencies operating with different sets of goals and assumptions (Cheers, 1995, 1998; Jones et al., 2001;

Maude, 2002). Where economic development has traditionally been the constituency of departments of industry and regional or state development, community as social development has often been left as a residual activity for housing, welfare and human service departments, which 'mop up' the poor and disadvantaged groups who have been bypassed in the rush for economic growth (Haughton, 1999). Although notable attempts to overcome this segmentation do exist, such as the Whitlam government's Australian Assistance Plan in the 1970s (Gleeson & Carmichael, 2001), in general governments have ignored calls by social policy analysts for a recoupling of economic and social policies (Jones et al., 2001).

It is in the area of rural policy that some headway has been made in formulating integrated programs of development. Once believed to be a 'farm' crisis, the rural downturn has had a major impact upon the economic, social and environmental viability of whole communities, while simultaneously eroding their ability to deal with problems in an effective manner. Economic and social policies, on their own, have done little to ameliorate the situation, prompting governments to seek alternative policy directions for tackling whole-of-community decline in rural towns and regions. Emerging in the mid to late 1990s, this new approach has been termed 'rural community development' (Day, 1998b; Cheers & Luloff, 2001) and combines the goals and methods of the community development approach with a uniquely rural program of development. This attempts to provide holistic or integrated solutions to the rural downturn that incorporate economic and social objectives. As will be discussed later, however, this goal has been only partially achieved.

In spite of the agreement among policy-makers and practitioners that community and economic development in rural areas need to be integrated, important differences of opinion still exist over the nature of the link between them and the level of priority that should be afforded to each. In the first and more instrumental interpretation, community development is seen as a necessary step in aiding rural businesses and communities to become more competitive in the global economy by building their skills, capacities and networks for economic advancement. Such thinking has been partly inspired by Robert Putnam's work on the importance of social capital building as a precondition for economic development (Putnam, 1993), and has been responsible for turning community development into the handmaiden of economic growth and global competition. In contrast, conventional community developers such as Cheers and Luloff (2001) see rural community development as involving attempts to build, strengthen and support rural communities – of which economic development is but one means. In other words, economic development becomes only one component of a much broader goal of community wellbeing (see also Labonte, 1999).

More recently, these approaches have been by-passed by the so-called 'triple bottom line' of economy, society and environment (Lawrence, 2003) in which economic development, environmental sustainability and social wellbeing are given equal priority within a single policy framework. This balanced approach to sustainability brings together two established but parallel trends towards integrating social objectives with natural resource management, on the one hand, and with local economic development, on the other. Sustainability encompassing social, economic and environmental aspects of rural areas has been given nominal support through the Commonwealth government's adoption of ecologically sustainable principles, but without a fully developed policy foundation (see Tonts, in chapter 11). The sentiments behind this approach are noble, yet it is shown later in this chapter that integration is difficult to achieve in practice, particularly within the current neoliberal policy framework where economic considerations still prevail (see also Lawrence, 2003).

The history of rural community development reveals that rural policy and programs in Australia have been built on shifting sands with little consistency in policy formulation or enactment. Turf wars have occurred (as in Western Australia) between different departments involved with – or displaced from responsibility for – overlapping issues (see also Tonts, in chapter 11), while the federal Coalition government initially attempted to shift responsibility for regional development to state and local government (Haslam-McKenzie, 2003a). It was only later – as a result of the rural backlash in the late 1990s – that the Commonwealth government was forced to re-evaluate its commitment to rural areas and develop new rural and regional programs, while remaining 'fundamentally ambivalent about its role in regional development' (Gerritsen, 2000: 133). Abrupt switches in administrative and policy arrangements appear to occur both in response to crises and also as a result of the desire of governments at federal and state levels to distinguish themselves from their predecessors (Beer, 2000). As a result of these frequent changes of tack, and the conflict between neoliberal principles of small government and rural needs, regional development provision is a patchwork of fragmented and poorly integrated programs.

CONTEMPORARY RURAL COMMUNITY DEVELOPMENT IN AUSTRALIA

Rural development projects without community involvement are today hard to find in a policy environment that privileges 'small government', 'partnerships' and 'self-help'. Government agricultural extension has largely been replaced by community-based group training and facilitation through Landcare and other initiatives, while empowerment and rights-based approaches have also fallen out of favour. The major emphasis in community development over the past decade has shifted from attempts, using government-funded professional workers, to empower and achieve equity for disadvantaged groups within rural localities, to place-based initiatives to counteract the disadvantage seen as experienced by whole rural communities. This assumption of homogeneity and common interests of the 'community' overlooks or downplays the social exclusion that may be experienced by some residents, and which is the focus of interest for many European programs to promote community wellbeing and tackle social exclusion (Edwards, 1998; Blaxter et al., 2003; Stevens et al., 2003). However, although an interest in social exclusion has slipped into the background in local community development in Australia, it still remains a concern for many rural support organisations, such as community and neighbourhood centres, carer support organisations and housing associations, and is part of the practice of many community care workers in rural areas. In addition, the electoral reversals attributed to a rural (and outer metropolitan) backlash in recent years have resulted in the return of some federal government programs for socially disadvantaged areas and 'at-risk' families, such as the federal government's Stronger Families and Communities Strategy (Munn & Munn, 2003), although these are small-scale compared with UK attempts to tackle social exclusion (Edgar, 2001). In Victoria too, after suffering from cutbacks and the damaging effects of competitive tendering under the neoliberal Kennett government, community development as an adjunct to community care is now experiencing something of a renaissance (Carter, 2000; Lynn, 2001).

CAPACITY BUILDING

Community capacity building has come into vogue in Australia in recent years as a key component of rural policy, eclipsing the earlier practices of community development to overcome disadvantage, on the one hand, and state sponsored regional development, on the other. The scale of environmental and social problems in rural areas, combined with the neoliberal emphasis on small government and individual and community self-reliance, have resulted in a search for alternative solutions. The success of Landcare and other community-based environmental programs directed attention towards 'community-level, self-help solutions involving "local" leadership and community capacity building' (Williams, 2003: 162). The ability to manage and respond to social, economic and environmental pressures was seen to depend on the 'capacity' or 'capability' of people and communities in rural and regional areas – that is, 'the ability, organisation, attitudes, skills and resources that communities have to improve their economic and social situation' (Cavaye, 1999: 1). The remedies proposed were self-help and, where necessary, the rebuilding of 'community' and 'social capital' as the basis for rural areas pursuing their own goals independently or in 'partnership' with government agencies.

Like community development, capacity building (or capacity development) had its origins in development efforts in the Third World from the 1980s. However, in Australia, there has been little attempt to recognise the importance of 'the larger policy-enabling environment' (UNDP, 1997: 13). Rather, Australian programs have shifted the focus from systems or policy environments to the community, and more particularly to actual or potential leaders within the community. While capacity development is consistently linked to empowerment by international organisations (Lusthaus et al., 1999), instrumental approaches predominate in Australia, particularly in local economic development, agricultural change management and natural resource management (for example, MRAP, 2001), and in public health (Hawe & Shiell, 2000; Whittaker & Banwell, 2002). Other capacity development principles that have received little support in Australia are the need for adequate resources and sufficient time for the desired 'learning and adaptation' (Bolger, 2000: 2) to occur. In fact, the Australian application of capacity building/development fits the warning given by Lusthaus et al. (1999: 2–3) that the concept has 'taken on many meanings and has been used as a slogan rather than as a term for rigorous development work'.

SOCIAL CAPITAL

Another concept that has been adopted with alacrity in Australia is 'social capital' (see also chapter 1). 'Building community' and 'building social capital' are expressions that are used interchangeably and are based on the belief that the decline in the vitality of local areas in the West is associated with a loss of 'community' and 'social capital', both of which are considered to contribute to economic development (Putnam, 1993). As Durlauf (2002: 259) argues: 'From the perspective of public social policy, social capital has been treated as a "missing link" in explaining the success or failure of different communities and even societies.'

The concept has been widely criticised as ambiguous, theoretically weak, and not backed by adequate empirical evidence (Durlauf, 2002; Portes, 1998; Whittaker & Banwell, 2002), and yet it has been adopted enthusiastically in community development programs in the United States, by the World Bank and in many countries, including Australia. Durlauf (2002: 272) considers that 'the popularity of social capital reflects a widespread belief that conventional economic approaches to behaviour seem inadequate for understanding problems such as the social pathologies of the inner city' – or, one might add, the problems of underdevelopment and rural decline. Woolcock and Narayan (2000), on the other hand, see the language of social capital as appealing to policy-makers accustomed to thinking in economic terms. More cynically, Portes (1998) attributes the concept's appeal to the relatively low cost of non-economic solutions to social problems.

LEADERSHIP

Capacity building in the international literature is seen as a learning approach, requiring a long period for capacities to be developed. In Australia, on the other hand, it has often been applied as a quick and relatively inexpensive remedy – either through provision of motivational events to 'inspire' members of a group assembled from the 'community' (and often seen to represent it) or through training directed at a small number of 'leaders' (or 'champions'). The assumption is that there will be a kind of trickle-down effect of skills, knowledge and motivation from leader to community. Thus, promoting community leadership is seen as a means to increase community capacity. For example, Andersen et al. (2001: 8) argue that:

> The leadership approach is based on a premise that individual development enhances community capacity ... It should also strengthen the community's capacity to identify opportunities and address crises in innovative ways.

A series of reports in Australia have 'argued that strengthening regional economies by encouraging local leadership ... [is] essential to the growth of the Australian economy' (Haslam-McKenzie, 2003b: 133).

However, local leadership has also been identified as problematic because of the tendency for rural leaders to be drawn from a narrow elite and for community initiatives to be driven (or perceived to be driven) by self-interest or a narrow definition of the community's interests. In one Victorian country town, Dibden and Cocklin (2003: 196) noted that a local development project was dominated by a narrow group of leaders 'drawn from a relatively wealthy sector of society composed of the bigger and more successful farmers and businessmen, i.e., those members of the community who are richest in a range of resources', as well as being 'relatively well connected to government and state-level non-government organisations'. Other community members felt that their interests and proposals had largely been ignored. Cheers and Luloff (2001: 140) see more participatory approaches as the answer to this problem, arguing that 'it is through broad-scale participation that we are most likely to expand the capacity and quality of community agency while limiting the role of entrenched elites'.

Leadership may be seen as characterised by dual attributes: as part of the human capital of a local community (see Tonts, in chapter 11) but also as contributing to the development of social capital. Through leadership training, the capacity of potential leaders – their skills and knowledge – is enhanced and the range of their contacts is expanded. However, rather than increasing collective social capital, leadership training may in fact build individual social capital, by helping those who have received the training to achieve personal advancement (such as taking up positions on boards of quangos or associations). There is the possibility too that they will be 'empowered by the program to change their lifestyle and move away from the community' (Haslam-McKenzie, 2003b: 145) rather than staying and contributing to community revitalisation, particularly if the community is seen as 'dying'. This view of the effects of leadership training conforms to Bourdieu's definition of social capital as a resource of individuals, used to advance their own interests and grounded in their class position (see also Dibden & Cocklin, 2003; Gray et al., 2002). Putnam (1993), on the other hand, sees social capital as restraining this tendency to self-interest.

Critical to achieving the participation of local people within government-supported community development is the identification of people as actual or potential leaders, and the kind of training provided. Haslam-McKenzie (2003b: 144), for example, presents Progress Rural WA as a project that emphasised participation and 'encouraged less well-known community "leaders" (such as older women active within their local communities) to be more prominent and to overtly share the "leadership" work roles'. In Australia, rural women's networks have promoted the role of women and provided training and support (Teather, 1992), resulting in increased recognition of the importance of women's contribution to, and leadership in, farming and rural communities (Alston, 1995, 1998; Garnaut et al., 1999). Women have been active in Landcare (Williams, 2003), and they have been courted by farming organisations anxious about declining membership (Dimopoulos & Sheridan, 2000: 31). Although rural and farming women still struggle to have their voices heard at the national level (Alston, 2000, and chapter 8), and many continue to work invisibly within their local communities, the validation of social and community development, and emphasis on participatory mechanisms, has opened a space for women to become more prominent in local development activities and to contest 'the purely economic dimension of community development by stressing the social implications' (Macgarvey & O'Toole, 2003: 193–94; see also Dibden & Cocklin, 2003). Moreover, women's leadership styles are seen as more compatible with the 'transformational leadership' now advocated. This is not to suggest that initiatives to enhance quality of life and community wellbeing, or to preserve the environment, are solely promoted or valued by women, but that they are compatible with the kind of localised, community-level activities that in the past have been seen as the domain of women.

SELF-HELP

From the mid-1990s, the previous emphasis of government agencies on economic and agricultural development has been broadened to encompass a more integrated approach, which recognises that rural communities are not synonymous with the farming sector, and that social objectives are also important to community viability. A welcome accompaniment to this change of direction has been:

> ... an increased willingness to recognise demands for more flexible responses to the needs of rural 'communities' – to be achieved by giving them the opportunity and capacity to tackle their own problems effectively. (Dibden, 2001: 7)

Many rural communities have embraced the notion of self-help and responded to decline through a variety of strategies, ranging from protest and lobbying against the threatened closure of local services to 'survival and revitalisation strategies' (Tonts, 2000: 66), which have enabled them to retain population and even to grow.

Self-help actions to develop local communities are encouraged by governments of both political persuasions, since they sit well with a wider policy agenda that promotes self-reliance and competitiveness (Tonts, 2000). These strategies have included encouraging existing and new industries and businesses, providing new services (often to replace those withdrawn), improving the appearance of the town, developing recreational facilities, attracting new settlers through housing schemes, and promoting tourism. A number of compilations have been made of examples of 'successful' rural communities and the lessons that may be drawn from their achievements (for example, Cahill, 1995; Collits, 2000; People Together Project, 2000; Rogers & Collins, 2001; Kenyon & Black, 2001). These are held up as models for other communities still struggling to escape from a downward spiral towards anticipated oblivion. Some of these community revitalisation projects have been entirely locally initiated and used only local resources, but many have been able to draw on financial support and advice from local, state and federal government agencies. While government funding is limited and not easy to obtain, it represents a useful resource for rural localities engaged in community action.

Some communities appear to have reversed the trend of decline that pervades much of rural Australia, prompting some authors to seek 'recipes' or principles of what constitute successful self-help initiatives. Kenyon and Black (2001: 25–26), for example, identify a number of themes emerging from their 14 case studies of 'positive local development', including a positive attitude, enthusiastic local champions and leadership, local entrepreneurialism, a smart use of outside resources and a youth focus. It is notable that, with the exception of 'smart use of outside resources', all these themes refer to endogenous processes. A problem with assessing the lessons emerging from the case studies in this and other collections is that accounts of 'successful' communities are generally based on self-reporting by members of the community, and tend to be characterised by 'boosterism' among a community wishing to promote itself favourably to the outside world or potential funders (Harvey, 1989: 7). Another problem with these success stories is that they tend to be lacking in detail or context, making it difficult to evaluate the weight to be given to internal factors as compared to other possible variables.

While the importance of local initiative should not be discounted, a certain level of local capacity is required to establish a community bank or similar enterprise. Not all rural towns will possess these capacities, since one of the outcomes of the withdrawal of services over recent decades has been depletion of their economic and human capital. Tonts (2000: 69) argues that 'without support from the higher tiers of government, local development initiatives can result in a highly ephemeral and uneven pattern of rural development'. In many examples of 'successful communities', individual leadership is underpinned by institutional or financial support from local, state or federal government, or from other organisations such as a local Chamber of Commerce. Cahill (1995), for instance, mentions the importance of well-connected local leaders, while two of Kenyon and Black's success stories have received support from the Deputy Prime Minister, who is also their local member of parliament. All of the eight groups studied by Cahill (1995: 19) had been successful in attracting outside funding for specific projects, some employed a facilitator and others had received in-kind support from local government or state government agencies. Among Kenyon and Black's (2001) 14 success stories, 11 had received on-going local government support and several had participated in state government community development programs. The one town without external support, Harrow in western Victoria, has developed a 'sound and light show' as a tourist attraction – a course of action that, depending as it does on its novelty value, is not widely replicable.

The communities that do not feature in the stories of success are the most disadvantaged, lacking established leaders, connections, submission-writing skills, marketable resources or evidence of the kind of initiative that attracts funding. Eversole (2003: 76) points to a tendency for local community development 'to overlook the extra-community relationships, which mediate a community's access to resources: the structures of disadvantage (geographic, educational, economic, political and so forth) that communities and their members face'. Because of the commitment of state and federal governments to economic rationalist approaches, which emphasise competitive efficiency over social equity, support for rural communities often fails to address the problems experienced by the most disadvantaged areas. Although government funding programs use the language of 'capacity building', they often require pre-existing capacity in order to apply. An example of the hurdles to be jumped in obtaining assistance is provided by the Commonwealth Regional Solutions Programme (DOTRS, 2000) for rural and regional communities facing economic challenges, with a declining population and/or high levels of unemployment and social disadvantage. The funding guidelines for this program specify that applicants must demonstrate a financial or in-kind contribution to the project, seek support from other sources, and possess organisational, financial and project management abilities. These requirements thus assume the prior existence of a level of social capital and community capacity that is likely to debar or discourage many disadvantaged communities.

Recent case studies exploring sustainability of rural towns (Cocklin & Alston 2003) lack the time depth that would enable a definitive assessment of whether particular communities are sustainable and why. However, they do make it clear that rural communities cannot be examined in isolation but as part of a wider region, and that their location (particularly their proximity to a city) and their built and natural environment influence their viability significantly. Studies in Australia by Hugo, Smailes and others demonstrate that towns that have attracted residents tend to be on the fringes of major cities, in densely-peopled regions, in coastal areas or in other favoured locations (Smailes & Hugo, 2003; Hugo, in chapter 4; Smailes, Griffin & Argent, in chapter 5). Similarly, Sorensen (2000) shows how geographical accessibility contributes to the prosperity of rural places. The 14 success stories outlined by Kenyon and Black (2001) include at least 11 towns that owed their success in part to possession of a tourist attraction, such as a natural spa, a striking natural feature (Wave Rock at Hyden, Western Australia), heritage buildings and a rich local history, or proximity to national parks or the coast. Moreover, these natural or historic features often also attract new settlers. In several instances, a flow of new (and often enterprising) residents, attracted by the natural amenities or heritage or location of the place, had preceded rather than followed community development action. This is not to say that the community initiatives launched were not beneficial, both in improving the quality of life of existing residents and attracting tourists and new settlers, but that community action may not be a necessary and sufficient condition for ensuring viability.

Some communities are simply more favourably placed to develop themselves than others – the ingredients for success are already present. Because of the influence of such factors, which are largely beyond the control of local people, the emphasis on capacity building and endogenous initiatives has been criticised on the grounds that it tends to 'blame the victim', by explaining the success of flourishing communities in terms of their capacity and motivation, which struggling communities are assumed to lack (Herbert-Cheshire, 2000). This has prompted Gray and Lawrence (2001b: 286) to criticise the tendency of governments to perceive the economic and social problems facing rural Australia 'in terms of individual characteristics rather than structural relationships' (see also Mowbray, 2000). Indeed, the success of local revitalisation efforts may be undermined by broader social processes. As Tonts (2000: 67–68) points out: (1) leadership, initiative and local resources are required but may not be available; (2) projects may be 'unsuccessful due to a lack of community motivation, poor leadership, or simply bad luck'; (3) community projects may be undermined as a result of 'changes in macroeconomic conditions or circumstances'; and (4) not all can succeed in a competitive environment. Both in obtaining funding and in promoting the community to the outside world, local community development has been seen as using the language of 'community' and 'collaboration' but pitting communities against each other in a climate of competition, in which only some can be winners:

> ... it is difficult to see how local development strategies can be successful if they are formulated in the context of a zero-sum competition with other communities targeting the same potential markets and investors. The outcome of this competition tends to be a pattern of uneven spatial development, since communities with stronger leadership, greater economic and physical resources and certain locational advantages tend to win at the expense of neighbouring communities. (Tonts, 2000: 67–68)

The outcome of such competition, as Gray and Lawrence (2001a: 111) point out, is that areas 'most in need of an economic boost may be least able to generate one'. One solution proposed (but vehemently rejected by country people) is voluntary euthanasia – the acceptance that some small towns are facing terminal decline and would be best assisted by 'a managed movement of population into larger viable, regional centres' (Forth, 2001: 81). This possibility receives covert support through government emphasis on self-sufficiency, which privileges some communities over others (Collits, 2000; Whittaker & Banwell, 2002: 258). Another possibility, as yet barely considered, is recognition by Australian governments of the value of country towns to their residents and to the farmers they support, as well as adoption of policies promoting land stewardship and multifunctional land uses enabling preservation of a sustainable rural landscape.

CONCLUSION

Since the mid-20th century when it first emerged, community development has been subject to change and contestation by its various proponents. Initially emerging under the hegemony of western capitalist growth, its commitment to self-help and empowerment was an instrumental one, more concerned with incorporating 'undeveloped' nations into the march of progress than with allowing them to determine their own destinies. While a radical and critical stream of community development emerged during the 1970s – and indeed still exists today – more recent manifestations of community development have a neoliberal flavour that, once more, emphasises the instrumental rather than empowering benefits of promoting self-help and self-reliance among impoverished and powerless populations. Nowhere is this more apparent than in rural Australia, where the rural downturn has impacted dramatically upon the economic and social wellbeing of entire communities.

In an era of small government, fiscal restraint and user-pays, the principle of empowering rural people to take responsibility for their own development is an attractive one that has been embraced wholeheartedly by governments of all political persuasions. Thus, we have seen the emergence of a range of programs for building social capital, enhancing individual and community capacity, nurturing new leaders or 'champions' of development, and promoting self-help for addressing rural community decline. Unfortunately, a critical review of the assumptions and practices underpinning these techniques reveals significant problems with contemporary approaches to rural community development. The 'winners' of this approach are more likely to be elite groups and individuals who possess the skills and contacts to compete for funding, or rural areas conveniently located close to metropolitan centres, natural amenities or tourist attractions. Those in greatest need, on the other hand, appear to be in a continuous race to the bottom. Similarly, the wellbeing of the rural economy continues to take precedence over that of the environmental and social aspects of rural community life in spite of the 'triple bottom line' rhetoric that prevails today.

In closing this chapter, it is worth considering, briefly, the future direction of rural community development in Australia, particularly in light of recent moves by the federal government away from the locality-based strategies of action that underpin the community development approach. It cannot be denied that small, localised development strategies pursued individually may be less viable than co-ordinated efforts to develop small towns as part of a wider sub-regional or regional system, which enables economies of scale and sharing of resources, risks, costs and benefits. Indeed, Gray and Lawrence (2001a) argue that regional or catchment-based development would provide the economies of scale needed to address issues of economic, social and environmental sustainability (see also Lawrence, 2003). The recently released Sustainable Regions Programme appears to respond to these concerns, offering the potential to bring communities together across regions to seek collaborative solutions to common problems, and to share resources, expertise and personnel. Whether the program actually achieves this aim, however, depends on the extent to which the regional committees entrusted with this funding are able to ensure maximum participation by a range of stakeholders, and to include, or at least consider, the interests of those who are currently excluded. Unfortunately, the composition of the regional advisory committees – drawn largely from local government and business (DOTRS, 2003) – suggests that regional development will experience the same problems of elite domination as local community development has in the past, as well as continuing to emphasise economic development at the expense of social and environmental sustainability. Any change to this trajectory is at present unlikely.

13

CHALLENGES FOR INDIVIDUAL AND COLLECTIVE ACTION

Ian Gray

Social scientists can find themselves confronted by two opposing, but both apparently reasonable, interpretations of sustainability. One interpretation is based on the observation that history is moving us all through an inevitable process of change over which we have no control. This lack of control is attributed to forces that we cannot, either singularly or collectively, influence. The idea that the process of globalisation will inevitably transform all our economies and societies is an example of this view. In the context of sustainability, the idea that we are headed for inevitable environmental disaster is also such a view. In contrast, some people believe that globalisation is not inevitable, or that we are making significant and effective changes to ensure that our environment will continue to support life as we know it. The future of small regional communities can also be seen from similarly opposing perspectives. From one perspective, it can be argued that contemporary economic, social and environmental systems are all steering our communities inevitably towards their demise. From the other perspective, it appears that people can take action to change the trajectory of decline. These perspectives have recently collided.

The collision occurred when Forth (2001), and later Forth and Howell (2002), attempted to counter an action-oriented approach that has almost become an orthodoxy among regional policy advocates. Forth's 2001 paper created a furore when delivered to a conference in Bendigo. The ensuing debate has raised policy issues at the same time as it has indicated a range of interpretations from the deterministic and

fatalistic to the active and optimistic. Forth and Howell argue, deter-
ministically, that for many small towns decline is inevitable. In their
view, it follows from this observation that government policy should
assist the dignified passing of those small towns which are bound to
disappear altogether. They contrast their view with that of the optimistic, action-oriented 'revivalists' who promote the idea that small
towns can save themselves. All they need do is galvanise into action.
This latter view has characterised much recent policy-making, having
clear foundations in neoliberalism (Gray & Lawrence, 2001a).

Forth and Howell argue, quite reasonably, that the claims of
'revivalists' are based on an unrealistic assessment of what small communities can achieve for themselves. The problem for the revivalists is,
according to Forth and Howell, the inevitability of decline in the
context of persistent population drift to cities and improvements in
transportation that facilitate population mobility. People can travel
more rapidly and conveniently to larger centres to conduct business,
which therefore attracts more business and hence population growth.
However, they also recognise that decline is not inevitable for some
small towns that might be advantaged by isolation. Given this point, it
is curious that they appear to assume that the development of transportation will continue and will persistently retain the same effects. This
is questionable, given that the standard of regional roads seems unlikely
to improve sufficiently to make reductions in travel times of the same
order that road sealing and improvement have enabled since the 1950s.

More importantly, they make the corresponding fundamental error
to that of the 'revivalists'. While the 'revivalists' assume that small-town
survival may be possible in all cases, Forth and Howell fail to acknowledge sufficiently that, while people believe that they can act against
decline, they may attempt to do so. Australian rural communities have
a tradition of interpreting their situation as requiring defence against
threatening, or at least unsupportive, interests. Denying them any possibility for success implies an assertion that their situation is totally exoge-
nously determined and cannot be escaped. For these reasons, the
argument that decline is inevitable is not an adequate basis either for
analysis of the forces acting with and against small regional communities, or for the design of policies and programs intended to help small-
town residents. At the same time, assertions that small towns can
'revive' themselves are not helpful either when they implicitly deny the
forces of history of the kind that Forth and Howell identify. What is
required is a way of thinking about the situation of rural communities

that acknowledges the extent of their predicament at the same time as it leaves the door open to community-based action.

When we acknowledge that regional communities can have, and do from time to time express, a collective interest (even though different interest groups may be identifiable within them) and recognise that there are forces against which such action must take place, we find people in challenging circumstances. This chapter investigates those challenges. It analyses them with one eye to objective analysis and the other eye to interpretations of those challenging circumstances by the people who confront them, the residents of small regional communities.

FROM 'CAPITALS' TO SUSTAINABLE DEVELOPMENT

Studies undertaken for the Academy of the Social Sciences in Australia project approached the notion of capital, in its five forms, as a resource that communities might or might not possess in smaller or larger quantities and which, if they have sufficient, will enable them to achieve sustainability (Cocklin & Alston, 2003). However, possession of capital is not an end in itself. Nor does it have any intrinsic value in supporting sustainability. Capital is only of value or use insofar as it can be, and is, applied to the problem of sustainability. To approach the question of application, it is necessary to think beyond the capitals concept in isolation.

Rather than see a capital only as a stock of a resource without any essential purpose, we can look upon it as something to transform into a practical asset that can be put to use by community members. The present condition of the stocks of various capitals is shaped by human action, and the capitals in turn constrain as well as enable action to promote sustainability. For example, natural capital may have been affected by many years of exploitation and may now be depleted to the extent that it is no longer available to the local economy. Social capital may have been eroded by inequality and exclusion, or it may be growing with local efforts to grow community organisations. The capitals are available to people to use, but sometimes they may be inadequate or get in the way. On the other hand, they may also be turned into assets. Social capital may provide a basis for learning and developing collective solutions to problems of sustainability. In this sense the capitals form part of a structure that presents itself to people as they seek sustainable development. In addition, there may be many elements of social structure that intervene between people and the capitals, making it difficult to transform them into assets. For example, social inequality or differentiation can threaten social capital as people try to organise across community social divisions. The Academy of the Social Sciences in Australia project case studies (Cocklin & Alston, 2003) reveal many of these intervening factors, or even barriers, as they apply concepts such as social class, gender and social capital at the local level, as well as overarching structures of external institutions like government. The idea of structure differs from capital in that structures embody relationships among people, institutions and communities. The capitals plus people and relations among those people, and relations among people and the capitals, are what make up structures. This chapter looks at the ways in which structures present challenges to people as they strive for sustainability.

For rural communities, the structures that confront them are often seen in a negative way. This is understandable in the common circumstance where the stock of a capital has been depleted, and the process of depletion might be attributed to a structural relationship, such as that between local farmers and global markets. But a way forward towards sustainability is only likely to be found when constraining structures are challenged and enabling structures applied. In a sense, local cultures can constitute enabling structures, perhaps most obviously where, say, an indigenous culture becomes part of a community's economic capital. Culture is also enabling if it is associated with social capital and a willingness to use it. On the other hand, culture is related to structure in potentially negative ways, as when social differentiation is maintained by a culture in which social exclusion is legitimised.

Accepting the challenge laid down by constraining structures is a dynamic process. We must add a dynamic element to our notion of structure. The idea of a system in which people act with or against structures is useful in this context. It helps us to see what happens in relation to structures. A system is dynamic, consisting of people going about their lives in a context of interaction and relationships with each other and the elements of structure, including the capitals. Australian federalism presents itself to local communities as a structure within which they must operate and maintain or obtain resources as best they can. But the maintenance and acquisition of resources, such as grants for local works, occurs in a system where the local-central relationship takes its form through the actions of many people in its various component parts. The structure is not an absolute. It is created and maintained by people acting in systematic ways. This is also true of local communities: they are local social systems in which relationships provide the context where structures are maintained or challenged. In this chapter, I examine elements of structure and system through an exploration of processes of sustainability in a number of Australian rural communities (Cocklin & Alston, 2003). People confront structures through systems; let us see what confronts people, so that we might look for better ways ahead.

CHALLENGES FOR RURAL COMMUNITIES

STRUCTURES OF DEPENDENCY

Perhaps the defining characteristic shared among all regional communities is economic dependency and the system through which it is maintained and challenged. The notion of dependency has been applied to regional communities in many places at many times. It illuminates situations in which communities of people rely on resources over which they lack control, and moreover, are restricted in their capacity to determine their own circumstances and futures. It was first used with reference to regional communities by Vidich and Bensman (1960) in their study of a small town in the United States, looking at the ways in which the town was dependent on metropolitan centres. It has more recently been applied to communities whose economies are based on resource extraction, such as mining and timber towns (see, for example, Smith & Steel, 1995) because such towns are often dependent on one or a few companies that are controlled from elsewhere. The same notion can be applied to situations in which towns are dependent on large externally controlled industries, such as food or fibre processors.

Dependency theory may be applied more broadly, however, to the relationship between metropolitan and regional Australia. Such research has a longer history in the United States (see Martindale & Hanson, 1969), but Gammage (1986) and Gray (1991) provide examples of Australian rural communities that have been subservient to metropolitan-based decision-making throughout their histories. Like Dollery and Soul (2000), it is reasonable to see the domination of regional organisations as a significant feature of modern industrial societies in general. A structure of dependency presents formidable barriers to those seeking sustainability for small towns, especially as regional Australia is a historical creation of metropolitan Australia (see, for example, Davison, in chapter 3). Regional industries and communities were created in an image that assumes that what is good for the metropolises is good for the regions. This can be problematic, even when the metropolises act to look after the regions, because when they do so they apply their own interpretation of what is mutually beneficial. Productivism and resource exploitation have been driven by metropolitan interests as much as they have been carried out by rural people. They have grown in what can be seen as an urban image that casts a long shadow over the dominant approach to Australia's development (for discussion, see Gray & Lawrence, 2001a).

Towns can be dependent on other, usually larger, towns as well as metropolitan institutions. Amid these interdependencies, formerly local institutions become regional, with their headquarters and/or larger service centres elsewhere. The case studies in Cocklin and Alston (2003) illustrate this phenomenon. For example, the small towns of the Gilbert Valley, South Australia, are finding that the economic focus of their region is moving to larger towns outside the valley (Smailes & Hugo, 2003), while in Guyra, New South Wales, the shift towards off-farm work may be directing business and other activity towards the larger regional centre, where more employment and services are available (Stayner, 2003).

The systems within which decisions are made by governments and corporations are important to small towns. Government and corporate decisions can bring either despair or prosperity. Hence, local communities seek to intervene in decision-making processes. The importance of access to the machinery of politics at state and national levels has long been apparent (see the rather optimistic discussion in Nalson & Craig, 1989). The phenomenon of the 'marginal seat' – with a disproportionate influence within state or national politics – is well known and the importance of access to political leadership can be apparent (see the case studies of Guyra, by Stayner, 2003, and of Gulargambone in Kenyon & Black, 2001). On the negative side of this ledger, rural communities can seldom influence major policy changes, such as dairy deregulation, forestry management change or local government amalgamation (see Cocklin & Alston, 2003). They often lack the resources required for effective lobbying. This is not helped by the system of local government whereby local councils are sometimes treated like departments of state governments or subservient to the wishes of government departments. This situation is cemented in the lack of national constitutional recognition of local government, but perhaps more significantly in the creation of local councils by state legislation that prescribes their means of governance, such that the states can even terminate the existence of a local council, or replace its democratic government with state-controlled administrators. Both these unfortunate circumstances have occurred in the course of the amalgamation of local councils, often despite determined local opposition. The unsuccessful campaign by 'Tarra' residents when their Shire and two others were merged (Dibden & Cocklin, 2003) presents an example. The system within which communities attempt to influence corporations may not be so visible. Dependency on the corporate world can be manifest in other ways. Real estate values may be so low that town property is not sufficient to be considered an asset against which loans can be made. This highlights regional dependency on outside financial institutions. It poses a considerable challenge for small communities, some of which have responded by directing their investments locally. The problems of attracting savings to local investment in a small economy can become apparent, however. Even the establishment of a 'community bank' can depend on external support, as can attempts to move away from economic dependency on resource exploitation towards 'postmodern' development (such as tourism). For example, a project to revitalise Narrogin by redeveloping its streetscape relied on external funding from the Western Australian government (Tonts & Black, 2003). Similar examples are provided by Kenyon and Black (2001). New rural industries that show some potential to arrest environmental decline may need outside as well as local support.

Dependency presents a formidable structure and a system that maintains it, but it is never entirely uni-directional. Sometimes governments and corporations need local resources. Development draws on local labour markets, and co-operation with local authorities can smooth progress. Some towns have found ways to resist external forces, and adapt the system to their needs, without necessarily changing the structure. Guyra provides an interesting example of reaction to apparent dependency and powerlessness (Stayner, 2003). Although unable to prevent closure of its main industry, and suffering immediate and substantial economic problems, the people of Guyra attempted to rebuild their local economy. They achieved some successes, although in part remaining dependent on external support and not always able to retain the benefits in their community. Some significant industry has been retained and new industries attracted. Dependency is thus not necessarily the end of the story. Achieving sustainability will always remain difficult when the dependency embodied in the metropolitan–regional relationship makes it difficult for small communities to ensure that change really serves their interests. Taking the opportunities that are offered is a challenge in itself, but one that has to be accepted if sustainability is to be achieved.

LOCAL SOCIAL STRATIFICATION

Small towns have their own social systems. In an early discussion about local community studies, sociologist Margaret Stacey (1969) suggested that all communities, whether small towns or suburbs of large cities, should be regarded as local social systems. Her point is that local communities are worth studying for what they reveal about social relations based on social structures. Social structures exist within small towns, just as small towns exist in the context of broader political and economic relations. I will look at internal community structures before considering systems and cultures below.

Even quite small towns can exhibit social stratification. Studies, such as those by Gray (1991) and Wild (1978), have described social areas, providing maps to show how different parts of the towns are distinguished by social class. The social/spatial characteristics of the towns studied by Gray (1991) and Tonts and Black (2003) were largely a product of historical public housing policy. In both these towns, race provides an additional dimension to spatial class stratification, with some areas having concentrations of Aboriginal people due to historical government assimilation and settlement policies and to resistance to change by the local white population (see also Cowlishaw, 1988). Class and other social cleavages can, however, be significant without a necessary spatial dimension. Hence they can be important in quite small towns and villages. Dempsey (1990) was able to attribute significance to aspects of social inequality like class, but also gender and age, in a town of 2700 people, without emphasis on a spatial dimension.

The challenge for sustainability endeavours is not, initially at least, so much to change the structure of dependency, for it is determined by many factors exogenous to the local community. The problem encountered in local initiatives is rather to ensure that whatever change can be achieved under relations of dependency does not increase social differentiation based on stratification in ways that divide the community, or effectively create increasingly distinctive sub-communities (or neighbourhoods in some circumstances). The dangers of assuming a singular community of interest are indicated by the work of Scott et al. (2000) in New Zealand and Gray (1991) in Australia. In the latter study, the interests of some clearly identifiable groups were effectively ignored in local decision-making. In one instance, a lower middle class/working class part of the town suffered an air pollution problem for many years while this was rendered a non-issue in the local government council.

The nature of inequality will depend heavily on the history of local development. Towns in agricultural areas with a significant historical basis in extractive or other non-primary industry – such as the timber mill in Tumbarumba (Wilkinson et al., 2003) or the railways in Narrogin (Tonts & Black, 2003) – are likely to have a working class, even if only in retirement. Other towns where the economy has been more completely based on serving agricultural industries may have a more prominent middle class based on small business and various service activities (as observed in Dempsey, 1990), although there may be some processing industry maintaining, or formerly maintaining, a working class.

Those who would challenge dependency and act for 'revival' need to be aware of whose interests they are really representing. This requires a sensitivity to be possessed by all local institutions to the local social structure, which people continually create and recreate, and the system within which they are embedded. Given that local institutions would most likely have developed in the context of the same local social system that has maintained inequality, developing inclusive rather than exclusive organisations presents a challenge. Local government is the most central of local institutions (see Gray, 1991). It is often the only one that can effectively represent local interests in wider arenas. The social sustainability of the Gilbert Valley was threatened more by the loss of local self-government than by any other factor in the late 1990s (Smailes & Hugo, 2003). However, local government also has a tradition of elitism and narrowness in its agendas, despite the well-meaning intent of generations of local coun-cil members across rural Australia. The community studies of Wild (1978 and 1983) and Gray (1991) indicate histories of local council membership drawn narrowly from the local elite. Higgins (1998) argues that the farming elites in local government can be resilient through structural change in agriculture and change in local economies. This runs deeper than local electoral politics. It is embed-ded in the system of local government. While state governments can readily alter the structure of local government, and have sometimes proved willing to make radical change, challenging the system from below is more difficult. Other local organisations may be more flexi-ble, but the challenge of ensuring that they all provide equality of opportunity and develop broad-based participation is substantial. This is especially problematic, as local cultures can maintain a 'taken for granted' acceptance of inequality.

DEPENDENCY

Power, as observed in relations of dependency when a decision of a government or corporation affects many people, or more simply when one person influences the actions of another, is always exerted over people who are free to respond and attempt to change the power relationship. Absolute power, as in the total control exerted in a slavery relationship, is not really power at all (for a theoretical discussion see Foucault, 1983). Only free people are subjected to power, and being free, they can be expected to act. For example, 'Tarra' residents have attempted to alleviate their financial dependency on major banks, through local borrowing and investment (Dibden & Cocklin, 2003), and several towns described in Cocklin and Alston (2003) have attempted to ease their dependence on traditional industries. This is significant even if those attempts have been at least partly dependent on outside support. Nevertheless, there is evidence of a culture of dependency in small regional communities. Overcoming that culture presents a challenge.

It is important to note that this need not be in the form of people 'waiting for the cavalry', as it is sometimes expressed, describing people waiting and hoping for outside assistance for their industries and institutions (see, for example, Forth, 2001: 78). The idea that towns 'wait for the cavalry' stems from a misinterpretation of the history of support given to agriculture in order to establish and maintain a farm system based on families, which very efficiently provide a source of labour and continuity. Perceptions of the cost of providing services to small, relatively remote communities have probably also fuelled notions of small-town dependency specifically on government, rather than more broadly, disadvantageously and in a sense politically, on metropolitan institutions.

Rather than 'waiting for the cavalry', there is plenty of evidence of towns having pursued their own interests for a long time. Of necessity, this may consist partly or entirely of seeking outside support in the form of public or private investment. Witness the local economic development organisation described in Gray (1991), the many initiatives analysed in Cahill (1995) and the local natural resource management initiative, which long predates Landcare, described in Wrigley (1988). In some situations, however, there can be tension over pushes for change. A community's identification with particular industries and resources can mean that, with restructuring, if not closure, of those industries, people are forced into exploration of what appears to be foreign territory.

The culture of dependency can be seen in the form of apathy (not caring), fatalism, fear (see Panelli, 2001) or possibly lack of awareness of possibilities for action. In Canada, Epp (2001) has observed the 'de-skilling' of rural communities as people develop a cynicism accompanying a distaste for politics – a kind of politics of anti-politics. Belief that local affairs should be free of politics has been recognised as helping to protect the interests of elites (Gray, 1991) and maintaining conservatism in Australian local government (Halligan & Paris, 1984; Kenyon & Black, 2001). The challenge for communities and their leaders is to avoid cultures of apathy, fear (including wariness of politics), unawareness and cynicism. This is a difficult challenge because of dependency. It will always be difficult for dependent communities to obtain any degree of control over their destiny.

COMMUNITY AND DIFFERENTIATION

Structural differentiation within communities prompts questions about what constitutes a community. It would seem unlikely that community-based action will occur unless the members of the community share a sense of belonging to their community. When and how community members band together to take action in their perceived collective interest has been the subject of study. Some writers, like Tilly (1974), identify crisis response as a significant explanation/cause of community action, while other writers look at other internal or external factors for explanation. Suttles (1972) says that, while territoriality is significant, it is external forces that bind communities together. In Australia, Wild (1983) relates events surrounding effective community action after members perceived a threat from state government. The threat of economic decline following closure of the abattoir (a major source of employment) initially precipitated a strong reaction in Guyra, but there was a loss of energy over time (Stayner, 2003). Hall et al. (1984) look at internal factors including propinquity – relations developed by people who live in close proximity, sharing common class characteristics, and sharing bonds of kinship – while Metcalfe's (1988) Australian work shows how a community whose members share a common class interest can create a formidable force.

Part of the complexity in all these analyses arises because binding is temporal. Communities can be bound tightly as they act together, but bonds can wither if action decreases in intensity. Wild (1983) applied a concept of 'communion' to describe this sort of temporary situation, reserving community for more stable relations. Further complexity arises because social closeness is not necessarily a precursor to strong community action or communion. 'Weak ties' based on informal relationships can be significant too (Granovetter, 1972) and, as Zablocki (2000) notes, citing the work of Flache and Macy (1997), 'strong ties' – as created by the bonds of family life – can lead to group failure, and even a shared sense of belonging does not necessarily lead to strong collective action when other values and beliefs are not also shared. The only certainty is that this evidence suggests that there is no clear relationship between social capital and community action (a point also made by Bridger & Luloff, 2001) – and without some sense of belonging across a community, collective action would be unlikely to occur.

As suggested theoretically by Zablocki (2000), people may be members of a community by way of residence, but they can also be members of non-spatial communities of interest and at the same time have allegiances to other places. For example, in the Gilbert Valley, people indicated a sense of belonging to small communities, but often had business and other affiliations to larger towns (Smailes & Hugo, 2003). Nevertheless, local social capital can be strong. The analysis of social capital remains useful in confronting the challenge of community organisation. Each of the studies in Cocklin and Alston (2003) offers evidence of social capital. Despite loss of young people who might provide energy and continuity and middle-class volunteers and leaders, voluntary organisations are strong in all towns. It would appear that many organisations are accepting the challenges presented by population change. There is evidence of adaptation, as when football teams from two small towns in the Gilbert Valley merged, thereby creating a potentially stronger Valley identity (Smailes & Hugo, 2003). Social stratification and differentiation can threaten social capital and present challenges to those seeking to foster and retain identification with a community that might become a resource for collective action. Differential identification among residents in a town and its neighbourhood has often been observed; for example, between railway people and other residents or between farmers and townspeople (Cocklin & Alston, 2003). This form of differentiation has frequently been revealed in Australian community studies, including Dempsey (1990), Gray (1991) and Oxley (1978). It can persist despite ready identification of common interests, particularly between farms and town businesses.

But does structural differentiation necessarily interfere with social capital and community identity? It may not affect either. Oxley (1978) found that, while class and status stratification intrudes in work, domestic and private relations, egalitarianism prevails in voluntary organisations. On the other hand, social exclusion based on race, gender and even age can be significant (Dempsey, 1990), to the extent that people are effectively excluded from community life. At the same time, even those most effectively excluded can express a sense of attachment to their local community. It is here that our knowledge of the relationship between social capital and community organisation lets us down. The problem for community-based action is clear enough, however. While social interaction may break down some aspects of social differentiation, exclusion can still occur, raising the prospect that the interests of some groups will be ignored.

There is a converse to this problem (identified by Zablocki, 2000). Some people may deliberately exclude themselves in order to pursue their own interests, as against the interests of the collective, while retaining the advantage offered by the collective. This is the classic 'free rider' problem. The theory of social capital suggests that people may identify with and benefit from co-operative endeavours, but a rational action model would suggest that co-operative action will only be chosen by people anticipating benefits that are not otherwise available. At the individual level, this would seem an uncertain basis for local development.

A parallel to this problem can be identified by community members when one of them uses the stock of one or more 'capitals' to their own advantage. It indicates the inherent tension in collective action in a market system. Reading Kenyon and Black (2001), one finds instances of the prosperity of small towns increasing, in part at least, due to the entrepreneurialism of one or more individuals, but the extent to which the collective will necessarily profit from individual enterprises may be questionable. The same local initiative, to which many community members contribute, can look like either a community success or an individual success, depending on whose interests are seen to be served. Ensuring that collective action results in collective gain presents a challenge.

CHANGE

Population loss and ageing are apparent in many small communities in Australia. This compounds their decline, as government and commercial services are lost with falls in the threshold populations claimed necessary to justify service maintenance. All the towns studied in Cocklin and Alston (2003) appear to be suffering systematic decline, although Narrogin, which has become a 'sponge' town, has relatively bright demographic prospects (Tonts & Black, 2003). In some situations, people are coming to live in small towns while working in nearby larger towns. Thus, both Guyra and the Gilbert Valley have become attractive as dormitories (Stayner, 2003; Smailes & Hugo, 2003). This brings additional population and people with fresh ideas, but it can also have adverse outcomes, such as increased rents and a proportionately small increase in trade for local businesses due to the leakage of business to larger towns. One of the challenges for small towns lies in taking greatest advantage of what positive factors the social system offers.

CONFLICT

Challenging decline and dependency is made difficult by systemic sources of conflict. Environmental and economic capital can provide grounds for conflict, when resources offered by one are seen to deny resources offered by the other. As Ramp and Koc (2001) show, conflict over resource exploitation is something that communities face under globalisation investment scenarios, and which may become part of local politics. As Molotch (1976) observes for the United States, the absence of any such response to development may indicate the presence of an elite grouping of business (basically property investment) interests and local elected representatives and officials, smothering potential issues for the sake of its own interpretation of development. Where issues arise, they may be manifestations of a lack of co-ordination and 'pulling together' among local organisations, all intent on development (Herbert-Cheshire & Lawrence, 2003).

We have seen how hard it is to work out, in spatial terms, where communities begin and end. The placement of administrative boundaries can create 'communities', in the sense that the people living within the boundaries share important institutions. But there may be 'natural' divisions within those communities, creating grounds for conflict. The challenge is to ensure that local institutions are able to recognise all interests and provide an arena in which issues can be resolved. The planning role of local government, as well as its size and centrality as a local institution, makes it the most likely arena. However, its traditions, particularly in country towns, are more attuned to its role as a development corporation, a 'people's corporation' as Gray (1991: 60) describes it, where a traditional kind of elitism is preferred to democracy. Some recent changes to local government in some states have sacrificed elements of democracy for the sake of economic efficiency (see Newnham & Winston, 1997).

As the leading local organisation and the only one that can effectively represent local interests in wider arenas, the resolution of conflict often falls to local government. However, local government suffers two institutional weaknesses as a political arena, in addition to its non-democratic traditions and possible futures under neoliberal reform. These weaknesses are, firstly, that it can find itself in conflict with the state level of government, on which it is substantially dependent and in many ways subservient, and secondly that it has limited capacity to foster social capital. The latter point is discussed by Worthington and Dollery (2000). They find that, while there are differences among the states, local government's narrow range of instrumental functions gives limited scope to foster a sense of place among residents. As Gray (1991) argues, local government also suffers from a policing role, such as the impounding of animals, and its main local source of revenue (rates) is a particularly blunt form of taxation. Developing a sense of attachment and organisational energy through local government is problematic though not necessarily impossible (see also Tonts, in chapter 11).

The possibility of conflict with state government is perhaps more significant, as it illuminates a double-bind that many local organisations can face when confronting their dependency. If the community is seen to be complaining about its predicament, state and federal authorities may feel that they are being blamed and may even feel affronted (as was demonstrated for the community of Monto by Herbert-Cheshire & Lawrence, 2003). Yet, if communities do not express their concerns, they have little prospect of obtaining the support that they need. This presents a challenge, calling for particularly delicate politicking. Maintaining 'bridging social capital' and using it sustainably are both important challenges. They lie on top of the problems that communities face in gaining access to the state and federal government bureaucratic systems through which they might hope to obtain support, such as funding, help with planning, employment of development staff, and so on.

An equally significant double-bind can arise when a dependent community is 'made an offer it cannot refuse'. Significant threats to sustainability may emerge when, for economic reasons, small towns feel that they cannot reject industries that might be of dubious social and environmental value. They find themselves locked into expensive competition with other towns. The winners may pay dearly for victory and a chance of further development for some indefinite but potentially finite period of time. The losers can only hope for a win at another time.

The air-polluting industry described by Gray (1991) is a good illustration of an expensive win of this kind.

Finding a pathway toward sustainable development is particularly hazardous in these circumstances. Dependency means that the fundamental tensions of sustainable development confront the most vulnerable communities, which lack both the political means to resolve them and significant prospects of finding alternatives. This is perhaps the greatest challenge facing the people who are striving to find sustainability in small Australian country towns. But it is a challenge that many townspeople and other rural dwellers appear willing to accept, even at some personal cost.

CONCLUSION

What are the prospects for small towns? Will many of them inevitably disappear? The elements of structure, culture and system as currently constituted appear to present formidable barriers. But there are enabling resources as well as formidable constraints. There is also evidence indicating that, despite dependency, resistance happens. Double-bind situations, however, make it difficult to think optimistically about the prospects for many communities. The stoutest resistance is trapped in such circumstances.

What, then, might be done by those, including governments, who are concerned about regional community decline and the increasing socioeconomic gap between metropolitan and regional (especially inland) Australia? It is possible to consider a range of policies, from those that are akin to the 'revivalist' model and attempt to help individuals to make the difference that will rescue their town, to those policies that seek to create the circumstances in which small towns can survive and prosper. The former policies include those aimed at leadership development, for example. The latter include provision of infrastructure or other forms of industry support. The former fits the revivalist model, while the latter implies acceptance that small-town decline will continue if circumstances are not changed. Between these two positions lies the possibility that policy might assist communities to meet challenges by both building their capacity for independent action and altering the terms of dependency. This would include action that increases the regional infrastructure base. Such a policy would have to be built on an appreciation of the challenges being faced: the structural, cultural and systemic. Awareness of these challenges and preparedness to confront them require a capacity among local people for introspection into their own community, as well as analysis of the 'big picture' for the structural context of dependency and all its accoutrements. Policy-makers would also benefit from sharing this knowledge.

ACKNOWLEDGMENTS

I am grateful to Lynda Cheshire for helpful comments on an early draft of this chapter, and to Jacqui Dibden and Chris Cocklin for their advice.

14

CONCLUSION

Chris Cocklin and Jacqui Dibden

In chapter 3, Graeme Davison alerts us to the fact that the malaise afflicting many of Australia's rural communities is deeply rooted in history. The processes of change that are talked about today have been in train for many decades – technological change, out-migration, loss of markets, rising indebtedness and so forth. What distinguishes the present is the combined force with which these have come to bear and the fact that preserving rural society no longer carries the political weight it once did. Processes of change over the last 100 and more years inevitably mean that our contemporary rural places are very different from those of the past. Those of the future will be different again.

Looking forward, we can imagine many different prospects for rural Australia. In partnership with groups of landholders, we recently developed and evaluated a range of possible futures for rural land use and communities (Cocklin et al., 2003). These hypothetical futures included a relatively bleak prospect, featuring diminishing returns to farming, the increasing subsumption of the family farm, declining rural populations, and increased degradation of the natural environment. Under these conditions, farmers might be risk-averse, integrated closely with their suppliers and purchasers, and very focused on making ends meet financially. The more optimistic scenarios described rural futures featuring improved farming conditions based on new technologies and management systems, increased spending by governments on research and development, a focus on long-term productivity, revitalisation of rural communities, and improved environmental outcomes. With better economic returns, farmers might be more inclined to pursue higher-risk strategies (for example, new niche products) and they might also be in a better position to retire marginal land from production.

Asked to choose between these ruralities, it is not surprising that the farmers and other community members participating in these discussions opted for a future that we described as 'smart and environmentally sustainable' (Cocklin et al., 2003: 24). Their support for a brighter future was, of course, founded partly on the suggestions within the scenarios of improved economic returns to farming. But there was a great deal more to it than this. Investments in new technologies and farming methods, a wider social acknowledgment of the need for improved environmental performance (and a willingness among the public to pay for it), and the suggestions of more vibrant and viable rural communities were widely pointed to as features of their preferred futures.

An interesting aspect of these conversations was the fact that not only did the landholders prefer the more optimistic futures, but they also considered them to be more likely. Reflecting on the bleaker scenarios, there was a view that they were akin to looking back. In contrast to this, they could envisage a 'self-reinforcing system, driven by new technology and better business operations, creating improved returns that are reinvested in the system' (Cocklin et al., 2003: 20). It would also be a future marked by a 'cultural shift towards recognition of the mutual benefits of making a profit and looking after the environment' (Cocklin et al., 2003: 20).

In a broad sense, the chapters in this book seek to offer other assessments of the future prospects for Australia's rural communities. Consistently, the contributors share, implicitly or explicitly, a preference for our rural places to have economically productive farms that impose a smaller footprint on our natural environment and that are part of viable communities and social networks. There is also a strong sense within these essays that, based on present trends, these aspirations might not be met, despite the optimism of the landholders in our study. There is evidence of a developing incompatibility between deregulated, competitive, intensive agriculture and a widening environmental crisis that threatens the productivity of agriculture as well as the health of rural towns and natural ecosystems. This environmental and social 'crisis' has forced a reappraisal of policy directions grounded in neoliberal understandings and discourse. While continuing to promote neoliberal principles, there are signs of a shift by governments at both state and federal levels towards acceptance that farmers cannot be expected to bear the burden alone for 'public good' environmental work. At the same time, there seems to have been an acceptance of the fact that social capital may need to be energised, and that it is not sufficient to assume that there are leaders in every rural community available and ready to chart the way towards a more prosperous future (cf., Rees & Fischer, 2002; for contrasting views, see chapters 11 and 12).

In the United Kingdom and the European Union, the international 'crisis of overproduction' in the 1970s (see, for example, Argent, 2002) has led to policies discouraging agricultural surpluses and promoting conservation and rural development goals. This represented a significant shift from state sponsorship of the intensification, concentration and specialisation in production which, according to Ilbery and Bowler (1998), characterised 'productivism'. The changing policy environment in the United Kingdom and Europe has been associated with a postulated transition to a 'post-productivist' agricultural regime (Ilbery & Bowler, 1998; Evans et al., 2002; Wilson, 2001; see also Argent, 2002, and Holmes, 2002, for a discussion of the relevance of this concept to Australia). The concept of post-productivism implies a change in agricultural regimes in advanced economies from a concern solely with the production of food and fibre, as in the so-called 'productivist era', to a rural regime that comprises a multitude of functions. This multifunctional character of the agricultural landscape is not only the subject of academic study, but has also been valued by politicians and the public in Europe. Thus, the Dutch Minister of Agriculture, Nature Management and Fisheries, spoke of the need for both government and the agrifood industry to sustain 'a vital countryside, a viable and diverse natural environment, landscapes that reflect our cultural identity, as well as the economy and welfare of our rural communities' (Apotheker, 2000: 11). 'Multifunctionality' – the idea that agriculture has environmental and social as well as economic functions – has become a cornerstone of European Union agricultural and trade policy (Potter & Burney, 2002; Hollander, 2004). In the United Kingdom, the World Trade Organisation negotiations have triggered a debate 'about how far and in which way to sustain a multifunctional agriculture and an agrarian policy agenda under increasingly liberalised market regimes' (Potter & Burney, 2002: 46). Meanwhile, Australia has been grappling with the opposite problem – how to combine an already liberalised economy with the need to move towards more sustainable land management and how to maintain viable rural communities.

Australian agricultural and natural resource policy has continued to be characterised by a commitment to the tenets of neoliberalism, but these exist in uneasy juxtaposition with a growing recognition of environmental and social vulnerability. A dilemma for Australian governments is the perceived incompatibility of providing support for farmers (to help them deal with environmental problems arising from farming activities) with the view that is held by Australia and other Cairns Group nations that this would constitute a thinly disguised non-tariff barrier, and would therefore be contrary to World Trade Organisation rules (Potter & Burney, 2002; Wynen, 2002).

Australian agriculture is clearly in transition – a transition that may signal a complete 'post-productivist' withdrawal from agriculture in some parts of Australia (Holmes, 2002), and new 'post-productivist' emphases away from the production of food and fibre on some farms (Argent, 2002), possibly towards a multifunctional countryside that focuses increasingly on food quality instead of quantity, and recognises the role of landholders in environmental protection and remediation of occupied landscapes (Cocklin et al., 2003). As Richard Stayner commented in chapter 7:

> … rural regions are now the setting for the expression of a much wider range of aspirations and values than was contained within the historical vision of 'nation building' via the extraction or modification of natural capital.

Establishing the necessary conditions for farmers to provide ecosystem services as well as being producers of food and fibre will necessitate new kinds of policy settings, and there are positive signs of some imaginative thinking in this respect (see, for example, Cocklin et al., 2003; VCMC & DSE, 2003; Young et al., 2003).

Another theme that resonates strongly in this book is the need to be both more creative and more inclusive when it comes to nurturing social capital. It is eminently clear that social capital underpins community viability, the effective stewardship of natural resources, and the productivity of primary production in the most fundamental ways. Smailes and Hugo (2003: 98) claimed that 'social capital is the most important influence on the sustainability of rural communities'. There is a predilection among Australian governments for the 'self-help' approach to community development, but Jacqui Dibden and Lynda Cheshire, in chapter 12, are of the view that this often favours established elites and is typically directed at economic rather than environmental and social issues. In chapter 13, Ian Gray points to a need to build the capacity of communities for independent action and to alter the terms of dependency. We are alerted in this book also to the fact that, if sustainability connotes equity, as indeed it should, then steps will have to be taken to redress the marginalisation of the 'other' in rural communities. In chapters 8 and 9, Margaret Alston commends policy initiatives that enhance participation, provide greater access to services, and draw the 'other' actively into the lives of their communities and society.

In developing visions of rural sustainability, it will also be necessary to contemplate new prospects for the geography of communities and settlements. Peter Smailes and colleagues, in chapter 5, and Smailes and Hugo (2003), argue that sustainability needs to be pursued at a level that sits somewhere between the small town (or local community) and the region. In the Gilbert Valley, it was found that 'community life may be thought of as having been sustained through a natural process of amalgamation and readjustment of local identity to a broader spatial scale' (Smailes & Hugo, 2003: 99). Dibden and Cheshire, in chapter 12, point out that:

> ... localised development strategies pursued individually may be less viable than co-ordinated efforts to develop small towns as part of a wider sub-regional or regional system, which enables economies of scale and sharing of resources, risks, costs and benefits.

Broadly similar thinking underpins Richard Stayner's comments, in chapter 7, about industry clusters as a means of fostering rural economic development, though, as he notes, the sparse human geography of much of rural Australia may make this difficult in practice.

A more sustainable future for rural Australia will also demand of policy-makers that the gross inconsistencies in policy settings be progressively redressed. While there is frequent reference by governments to 'triple bottom line' outcomes, the reality is that there is little evidence of integration of policy around sustainability, leading Matthew Tonts to remark, in chapter 11, that 'one of the key challenges for governments in the 21st century is to begin to address issues of sustainability in a genuinely integrated way'. Establishing the balance of responsibilities in driving a rural sustainability agenda will require ongoing attention. There is a tendency to attribute the task to government, yet there is little evidence to suggest that this would be effective and, in any event, governments lack the resources to undertake the tasks alone. The investment of private capital on a substantial scale is required.

Governments can facilitate this, but ultimately the rates of return must be sufficient for investors to risk their hand.

The possible advent of carbon trading is one example of how this might happen, but also of the lack of consistency in government environmental policy. A carbon trading scheme would open up the prospect of substantial new investment in land uses (principally forestry) that could provide a new income stream to farmers, with ensuing community benefits (though we acknowledge that there are downsides as well: see, for example, Barlow & Cocklin, 2003). However, the future of carbon trading has been cast in doubt by the federal government's refusal to sign the Kyoto Protocol, and the recent halting of work on an emissions trading scheme (Peatling & Riley, 2004) in favour of 'a technologically based climate change strategy' (O'Loughlin, 2004). Other systems of financial reward to rural landholders for the provision of ecosystem services also indicate a potential for new kinds of private investment, but these also rely on a favourable policy framework, including taxation incentives (Sammon & Thomson, 2003). Of course, rural communities themselves must play a central part in charting their own destiny but, as we have indicated above, it is not sufficient to rely simply on a 'self-help' strategy.

As the title of this book is intended to convey, sustainability and change go hand in hand. If our benchmark is the status quo, sustainability is unattainable. There is no conceivable prospect that rural landscapes can be preserved in something like their current forms, and in many respects, neither should they be (for example, because of wasteful patterns of natural resource use). It will not be possible to secure a future for every small town, the demography of rural Australia will change, some rural settlements will thrive and grow, new agricultural commodities will be produced, there will be a shift towards the increased consumption of rural amenity, some natural environments will be irreversibly degraded while others will be restored, and values and cultures will change. The challenge to policy-makers, private capital and rural communities will be to steer a path through the inexorable forces of change in a direction that can be judged, widely, as a progression towards sustainability. The contributors to this book have marked out many of the important patterns of change and have presented some suggestions as to what might be done in the interests of aiming towards more sustainable rural futures. There can be no doubt that achieving change in support of rural sustainability will require a great deal of effort, as well as an openness to ruralities that are quite different from those we are familiar with.

REFERENCES

Addison, SC (2001) *Policy and Practice of Implementing LA21 at the Local Level in New South Wales and New Zealand*. Unpublished Masters degree thesis, Macquarie University, Sydney.

Agriculture, Fisheries and Forestry Australia (AFFA) (1999) *Managing Natural Resources in Rural Australia for a Sustainable Future: A Discussion Paper for Developing a National Policy*. AFFA, Canberra.

Aitkin, D (1988) 'Countrymindedness' – The spread of an idea. In SL Goldberg & FB Smith (eds.) *Australian Cultural History*. Cambridge University Press, Melbourne, pp. 50-58.

Allen, P (ed.) (1993) *Food for the Future: Conditions and Contradictions of Sustainability*. John Wiley, New York.

Almas, R & Lawrence, G (eds.) (2003) *Globalization, Localization and Sustainable Livelihoods*. Ashgate, Aldershot.

Alston, M (1995) *Women on the Land: The Hidden Heart of Rural Australia*. UNSW Press, Sydney.

—— (1996) *Goals for Women: Improving Media Representations of Women's Sport. A Report to the Department for Women*. NSW Centre for Rural Social Research, Charles Sturt University, Wagga Wagga.

—— (1997) Violence against women in a rural context. *Australian Social Work* 50(1): 15–22.

—— (1998) 'There are just no women out there': How the industry justifies the exclusion of women from agricultural leadership. *Rural Society* 8(3): 187–208.

—— (2000) *Breaking Through the Grass Ceiling: Women, Power and Leadership in Rural Organisations*. Harwood Publishers, England.

—— (2002) Inland rural towns: Are they sustainable? Keynote address to the Australian Bureau of Agricultural and Resource Economics (ABARE) conference, Canberra, March.

—— (n.d.) Gender and carework in Australia. Unpublished paper. Centre for Rural Social Research, Charles Sturt University, Wagga Wagga.

Alston, M & Kent, J (2001) *Generation X-pendable: Young, Rural and Looking for Work. An Examination of Young People's Perceptions of Employment Opportunities in Rural Areas*. Centre for Rural Social Research, Charles Sturt University, Wagga Wagga.

—— (2004) *Social Impacts of Drought: A Report to the NSW Department of Agriculture and the NSW Premiers Department*. Centre for Rural Social Research, Charles Sturt University, Wagga Wagga.

Alston, M & McKinnon, J (2001) Introduction. In M Alston & J McKinnon (eds.) *Social Work: Fields of Practice*. Oxford University Press, Melbourne, pp. 3–15.

Alston, M, Pawar, M, Bell, K & Kent, A (2001a) *Tertiary Education Access in the Western Riverina. A Report to the Riverina Regional Development Board*. Centre for Rural Social Research, Charles Sturt University, Wagga Wagga.

Alston, M, Pawar, M, Bell, K, Kent, A & Blacklow, N (2001b) *The Western Riverina Higher Education Needs Analysis*. Centre for Rural Social Research, Charles Sturt University, Wagga Wagga.

Anderson, J (1999) *One Nation or Two? Securing a Future for Rural and Regional Australia*. Address by John Anderson, Deputy Prime Minister, to the National Press Club, Wednesday, 17 February 1999, Canberra. Department of Transport and Regional Services. At: <http://www.ministers.dotars.gov.au/ja/speeches/1999/as01_99.htm>.

Andersen, L, O'Loughlin, P & Salt, A (2001) *Community Leadership Programs for NSW*. Preliminary Audit for Strengthening Local Communities Unit, NSW Premier's Department. University of Technology, Sydney.

Apotheker, H (2000) Is Agriculture in Need of Ethics? *Journal of Agricultural and Environmental Ethics* 12: 9–16.

Appadurai, A (1990) Disjuncture and difference in the global cultural economy. In M Featherstone (ed.) *Nationalism, Globalization and Modernity*. Sage, London, pp. 295–310.

Argent, N (1996) The globalized agriculture-finance relations: South Australian farm families and communities in a new regulatory environment. In D Burch, R Rickson & G Lawrence (eds.) *Globalization and Agri-food Restructuring: Perspectives from the Australasia Region*. Avebury, London, pp. 283–300.

—— (2002) From pillar to post? In search of the post-productivist countryside in Australia. *Australian Geographer* 33(1): 97–114.

Argent, N & Rolley, F (2000) Lopping the branches: Bank branch closure and rural Australian communities. In B Pritchard & P McManus (eds.) *Land of Discontent: The Dynamics of Change in Rural and Regional Australia*. UNSW Press, Sydney, pp. 140–68.

Ashby, J & Midmore, P (1996) Human capacity building in rural areas: The importance of community development. In P Midmore & G Hughes (eds.) *Rural Wales: An Economic and Social Perspective*. Welsh Institute of Rural Studies, Aberystwyth, pp. 105–23.

Asheim, GB (1986) Hartwick's rule in open economies. *Canadian Journal of Economics* 19: 395–402.

Atkinson, A (1996) *The Europeans in Australia: A History*. Volume 1. Melbourne University Press, Melbourne.

Austin, AG (1961) *Australian Education, 1788–1900: Church, State and Public Education in Colonial Australia*. Pitman, Carlton.

Australian Bureau of Agricultural and Resource Economics (ABARE) (1997) Changing Structure of Farming. *ABARE Current Issues* 4: 1–7.

—— (1999) *Changes in Nonmetropolitan Population, Jobs and Industries: Preliminary Report to the Department of Transport and Regional Services*. Commonwealth of Australia, Canberra.

—— (2000) *Australian Dairy Industry 2000*. Commonwealth of Australia, Canberra.

Australian Bureau of Statistics (ABS) (1994) *Australian Social Trends 1994*. Australian Bureau of Statistics, Canberra.

—— (2001) *Australian Social Trends 2001*. Australian Bureau of Statistics, Canberra.

—— (2002a) *Australian Standard Geographical Classification (ASGC)*. Australian Bureau of Statistics, Canberra.

—— (2002b) *Deaths Australia 2001*. Catalogue No. 3302.0. Australian Bureau of Statistics, Canberra.

—— (2002c) *Population Distribution, Aboriginal and Torres Strait Islander Australians* (2001). Commonwealth of Australia, Canberra.

—— (2002d) *Regional Population Growth, Australia and New Zealand*. Australian Bureau of Statistics, Canberra.

—— (2003a) *Australian Social Trends 2003*. Catalogue No. 4102.0. Australian Bureau of Statistics, Canberra.

—— (2003b) *Deaths Australia 2002*. Catalogue No. 3302.0. Australian Bureau of Statistics, Canberra.

—— (2003c) *The Health and Welfare of Australia's Aboriginal and Torres Strait Islander Peoples* (2003). Commonwealth of Australia, Canberra.

—— (2003d) *Population Characteristics, Aboriginal and Torres Strait Islander Australians* (2001). Commonwealth of Australia, Canberra.

—— (2004) *Measuring Social Capital: An Australian Framework and Indicators.* Commonwealth of Australia, Canberra.

Australian Bureau of Statistics (ABS) & MapInfo Pty Ltd (2002) *CData 2001.* Australian Bureau of Statistics, Belconnen.

Australian Bureau of Statistics (ABS) & Office of the Status of Women (OSW) (1995) *Australian Women's Year Book 1995.* AGPS, Canberra.

Australian Bureau of Statistics (ABS) & Space-Time Research Pty Ltd (1990) *1981/86 Census.* Australian Bureau of Statistics, Belconnen.

Australian Council of Social Service (ACOSS) (2001) *New research: Hardship intensifies from rise in social security penalties.* Press release, 13 August. At: <http://www.acoss.org.au/media/2001/mr0813.htm>.

Australian Institute of Health and Welfare (2002) *Australia's Health 2002. The 8th Biennial Health Report of the Australian Institute of Health and Welfare.* Commonwealth of Australia, Canberra.

Australian Local Government Women's Association (ALGWA) (2001) *National Framework for Women in Local Government.* Department of Transport and Regional Services, Office of the Status of Women and the Australian Local Government Association. Canberra.

Australian National Audit Office (ANAO) (1997) *Commonwealth Natural Resource Management and Environment Programs: Australia's Land, Water and Vegetation Resources.* AGPS.

Australian State of the Environment Committee (2001) *Australia State of the Environment 2001.* Independent Report to the Commonwealth Minister for the Environment and Heritage. CSIRO Publishing, Canberra.

Badcock, B (1997) Recently observed polarising tendencies in Australian cities. *Australian Geographical Studies* 35(3): 243–59.

Bailey, P & James, K (2000) *Riverine and Wetland Salinity Impacts – Assessment of Research and Development Needs.* Land and Water Resources Research and Development, Canberra.

Bamford, E & Dunne, L (1999) Quantifying access to health facilities: Leaping the boundary fence. *Proceedings of 5th National Rural Health Conference, Adelaide.* National Rural Health Alliance, Deakin West, ACT.

Bamford, EJ, Dunne, L, Hugo, GJ, Taylor, DS, Symon, BG & Wilkinson, D (1999) Accessibility to general practitioners in rural South Australia. *The Medical Journal of Australia* 171(11/12): 614–16.

Barbier, E (1987) The concept of sustainable economic development. *Environmental Conservation* 14: 101–10.

Barlow, K & Cocklin, C (2003) Reconstructing rurality and community: Plantation forestry in Victoria, Australia. *Journal of Rural Studies* 19(4): 503–19.

Barr, N (2000) *Structural Change in Australian Agriculture: Implications for Natural Resource Management.* Department of Natural Resources and Environment, Victoria.

—— (2002) *Victoria's Small Farms.* Centre for Land Protection Research, Report 10. Department of Natural Resources and Environment, Victoria.

Barr, N, Ridges, S, Anderson, N, Gray, I, Crockett, J, Watson, B & Hall, N (2000) *Adjusting for Catchment Management: Structural Adjustment and its Implications for Catchment Management in the Murray-Darling Basin.* Murray-Darling Basin Commission, Melbourne.

Bebbington, A & Perreault, T (1999) Social capital, development, and access to resources in highland Ecuador. *Economic Geography* 75(4): 395–418.

Beck, SL (1999) *Progress Towards Economy Indicators*. Sustainable Community Roundtable, Olympia, WA.

Beck, U (2000) *What is Globalization?* Polity Press, Cambridge.

Becker, RA (1982) Intergenerational equity: The capital-environment trade-off. *Journal of Environmental Economics and Management* 9: 165–85.

Beder, S (1996) *The Nature of Sustainable Development*. Second edition. Scribe Publications, Newham.

Beechey, V (1987) *Unequal Work*. Verso, London.

Beer, A (2000) Regional development and policy in Australia: Running out of solutions? In B Pritchard & P McManus (eds.) *Land of Discontent: The Dynamics of Change in Rural and Regional Australia*. UNSW Press, Sydney, pp. 169–94.

Beer, A & Maude, A (2002) Community development and the delivery of housing assistance in non-metropolitan Australia: A literature review and policy study. Positioning Paper for the Australian Housing and Urban Research Institute. January.

Beer, A, Maude, A & Pritchard, W (2003) *Developing Australia's Regions: Theory and Practice*. UNSW Press, Sydney.

Bell, J & Pandey, U (1989) Gender-role stereotypes in Australian farm advertising. *Media Information Australia* 51: 45–49.

Bell, M & Hugo, GJ (2000) Internal Migration in Australia 1991–1996, *Overview and the Australia-born*. AGPS, Canberra.

Beresford, Q, Bekle, H, Phillips, H & Mulcock, J (2001) *The Salinity Crisis: Landscapes, Communities and Politics*. University of Western Australia Press, Nedlands.

Berkes, F & Folke, C (1993) A systems perspective on the interrelationships between natural, human-made and cultural capital. *Ecological Economics* 5(1): 1–8.

Beus, C & Dunlap, R (1994) Endorsement of agrarian ideology and adherence to agricultural paradigms. *Rural Sociology* 59: 462–84.

Birrell, R (2000) Quoted in: Concerns over greying of rural NSW. *Sydney Morning Herald*, 22 March.

Black, A & Hughes, P (2001) *The Identification and Analysis of Indicators of Community Strength and Outcomes*. Department of Family and Community Services, Canberra.

Black, A & Kenyon, P (eds.) (2001) *Small Town Renewal: Overview and Case Studies*. Rural Industries Research and Development Corporation, Canberra.

Blainey, G (ed.) (1957) *Sir Samuel Wadham: Selected Essays with a Biographical Study*. S.M. Wadham Testimonial Trust Committee, Melbourne.

—— (1963) *The Rush That Never Ended: A History of Australian Mining*. Melbourne University Press, Melbourne.

Blaxter, L, Farnell, R & Watts, J (2003) Difference, ambiguity and the potential for learning: Local communities working in partnership with local government. *Community Development Journal* 38(2): 130–39.

Bolger, J (2000) *Capacity Development: Why, What and How*. Capacity Development Occasional Series, Canadian International Development Agency (CIDA).

Bolton, G (1981) *Spoils and Spoilers: Australians Make their Environment, 1788–1980*. George Allen & Unwin, Sydney.

—— (1994) *A Fine Country to Starve in*. University of Western Australia Press, Nedlands.

Bolton, R (1992) 'Place prosperity versus people prosperity': An old issue with a new angle. *Urban Studies* 29: 185–203.

Bonanno, A, Busch, L, Friedland, W, Gouveia, L & Mingione, E (eds.) (1994) From *Columbus to ConAgra: The Globalisation of Agriculture and Food*. University Press of Kansas, Kansas.

Boully, L (2001) *Leadership and Management Challenges for the Murray Darling Basin – Do We Have What it Takes?* At: <http://www.abc.net.au/4corners/water/boully.htm>.

Bourdieu, P (2001) *Masculine Domination*. Polity Press, Cambridge.

Bourke, L (2001a) One big happy family? Social problems in rural communities. In S Lockie & L Bourke (eds.) *Rurality Bites: The Social and Environmental Transformation of Rural Australia*. Pluto Press, Sydney, pp. 89–102.

—— (2001b) Rural communities. In S Lockie & L Bourke (eds.) *Rurality Bites: The Social and Environmental Transformation of Rural Australia*. Pluto Press, Sydney, pp. 118–28.

Bradshaw, B, Cocklin, C & Smit, B (1998) Subsidy removal and farm-level stewardship in Northland, New Zealand. *New Zealand Geographer* 54: 12–20.

Brady, EJ (1918) *Australia Unlimited*. G. Robertson, Melbourne.

Bray, JR & Mudd, W (1998) *The Contribution of DSS Payments to Regional Income*. Department of Social Security Technical Series Number 2. DSS, Canberra.

Bridger, JC & Luloff, AE (2001) Building the sustainable community: Is social capital the answer? *Sociological Inquiry* 71(4): 458–72.

Brown, N (1995) *Governing Prosperity: Social Change and Social Analysis in Australia in the 1950s*. Cambridge University Press, Melbourne.

Brunckhorst, D (2002) Insitutions to sustain ecological and social systems. *Ecological Management and Restoration* 3(2): 108–16.

Bryant, C (1992) *Agriculture in the City's Countryside*. Belhaven Press, London.

Bryant, C, Russwurm, L & McLellan, A (1982) *The City's Countryside: Land and Its Management in the Rural-Urban Fringe*. Longman, London.

Bryson, L & Mowbray, M (1981) Community: The spray-on solution. *Australian Journal of Social Issues* 16(4): 255–67.

Burch, D (forthcoming) *Transnational Agribusiness and Thai Agriculture: Globalisation and the Political Economy of Agri-food Restructuring*. Report to the Institute of South East Asian Studies, Singapore.

Burch, D, Lyons, K & Lawrence, G (2001) What do we mean by 'green'? Consumers, agriculture and the food industry. In S Lockie & B Pritchard (eds.) *Consuming Foods: Sustaining Environments*. Academic Press, Brisbane.

Burch, D & Rickson, R (2001) Industrialised agriculture: Agribusiness, input-dependency and vertical integration. In S Lockie & L Bourke (eds.) *Rurality Bites: The Social and Environmental Transformation of Rural Australia*. Pluto Press, Sydney, pp. 165–77.

Burch, D, Rickson, R & Annels, R (1992) The growth of agribusiness: Environmental and social implications of contract farming. In G Lawrence, F Vanclay & B Furze (eds.) *Agriculture, Environment and Society: Contemporary Issues far Australia*. Macmillan, Melbourne, pp. 259–77.

Burch, D, Rickson, R & Lawrence, G (eds.) (1996) *Globalization and Agri-food Restructuring: Perspectives from the Australasia Region*. Avebury, London.

Burchardt, T (2000) Social exclusion: Concepts and evidence. In D Gordon & P Townsend (eds.) *Breadline Europe: The Measurement of Poverty*. Policy Press, Bristol, pp. 385–406.

Bureau of Rural Sciences (BRS) (2000) *National Plantation Inventory Tabular Report*. BRS, Canberra.

Burkey, S (1993) *People First: A Guide to Self-Reliant, Participatory Rural Development*. Zed Books, London.

Burnley, I & Murphy, P (2003) *Sea Change: Movement from Metropolitan to Arcadian Australia*. UNSW Press, Sydney.

Butlin, NG (1964a) *Forming a Colonial Economy: Australia 1810–1850*. Cambridge University Press, Melbourne.

—— (1964b) *Investment in Australian Economic Development, 1861–1900*. Cambridge University Press, Cambridge.

Buttel, F (1994) Agricultural change, rural society, and the state in the late twentieth century: Some theoretical observations. In D Symes & A Jansen (eds.) *Agricultural Restructuring and Rural Change in Europe*. Agricultural University, Wageningen, pp. 13–31.

Cahill, G (1995) *Growing Your Own Community: Successful Adjustment Strategies for Rural Communities*. Agriculture Victoria, Bendigo.

Campbell, H (1996) Organic agriculture in New Zealand: Corporate greening, transnational corporations and sustainable agriculture. In D Burch, R Rickson, & G Lawrence (eds.) *Globalization and Agri-food Restructuring: Perspectives from the Australasia Region*. Avebury, London, pp. 153–69.

—— (2000) The glass phallus: Pub(lic) masculinity and drinking in rural New Zealand. *Rural Sociology* 65(4): 562–81.

Carney, D (ed.) (1998) *Sustainable Rural Livelihoods: What Contribution Can We Make?* Department of International Development, London.

Carr, A (2002) *Grass Roots and Green Tape: Principles and Practices of Environmental Stewardship*. Federation Press, Annandale.

Carter, J (2000) *Report of the Community Care Review*. Department of Human Services, Melbourne.

Castells, M (2000) *The Rise of the Network Society*. Second Edition. Blackwell, Oxford.

Cavaye, J (1999) *The Role of Government in Community Capacity Building*. Queensland Department of Primary Industries, Brisbane.

Chambers, R & Conway, GR (1992) *Sustainable Rural Livelihoods: Practical Concepts for the 21st Century*. Institute of Development Studies, Brighton.

Cheers, B (1995) Linking social and economic development in rural Australia. *Northern Radius* November: 14–16.

—— (1998) *Welfare Bushed: Social Care in Rural Australia*. Ashgate, Aldershot.

Cheers, B & Luloff, A (2001) Rural community development. In S Lockie & L Bourke (eds.) *Rurality Bites: The Social and Environmental Transformation of Rural Australia*. Pluto Press, Sydney, pp. 129–42.

Chiotti, Q (1995) The liberalisation of agricultural trade: Are we moving towards a more sustainable rural system? In C Bryant & C Marois (eds.) *The Sustainability of Rural Systems*. University of Montreal, Montreal.

Civil Service Commission 1858–59 (1859–60) *Victorian Parliamentary Papers* 2(19).

Cocklin, C (1993) What does cumulative environmental effects assessment have to do with sustainable development? *Canadian Journal of Regional Science* 16(3): 453–79.

—— (1995) Agriculture, society and environment: Discourses on sustainability. *International Journal of Sustainable Development and World Ecology* 2: 240–56.

Cocklin, C & Alston, M (eds.) (2003) *Community Sustainability in Rural Australia: A Question of Capital?* Centre for Rural Social Research, Charles Sturt University, Wagga Wagga.

Cocklin, C, Blunden, G & Moran, W (1997) Sustainability, spatial hierarchies and land-based production. In B Ilbery, Q Chiotti & T Rickard (eds.) *Agricultural Restructuring and Sustainability: A Geographical Perspective.* CAB International, Wallingford, pp. 25–39.

Cocklin, C & Dibden, J (2002a) Taking stock: Farmers' reflections on the deregulation of Australian dairying. *Australian Geographer* 33(1): 29–42.

—— (2002b) Deregulating the Australian dairy industry. In P Holland, F Stephenson & A Wearing (eds.) *2001, Geography – A Spatial Odyssey. Proceedings of the 3rd NZGS/IAG Joint Conference.* New Zealand Geographical Society, Hamilton, New Zealand, pp. 356–62.

Cocklin, C, Dibden, J & Mautner, N (2003) *Stewards of the Land: Landholder Perspectives on Sustainable Land Management.* Report prepared for the Victorian Catchment Management Council and Department of Sustainability and Environment. Department of Sustainability and Environment, Melbourne.

Cocklin, C & Wall, M (1997) Contested rural futures: New Zealand's East Coast Forestry Project. *Journal of Rural Studies* 13: 149–62.

Cohen, AP (1985) *The Symbolic Construction of Community.* Tavistock, London.

—— (1989) The Symbolic Construction of Community. Routledge, London.

Collits, P (2000). Small town decline and survival: Trends, success factors and policy issues. Paper presented at the First National Conference on the Future of Australia's Country Towns, Bendigo, Victoria, June. At: <http://www.regional.org.au/au/countrytowns/>.

—— (2001) Small town decline and survival: Trends, causes and policy issues. In MF Rogers & YMJ Collins (eds.) *The Future of Australian Country Towns.* Centre for Sustainable Rural Communities, La Trobe University, Bendigo, pp. 32–56.

Commins, P (1990) Restructuring agriculture in advanced societies: Transformation, crisis and responses. In T Marsden, P Lowe & S Whatmore (eds.) *Rural Restructuring: Global Processes and their Responses.* David Fulton, London, pp. 45–76.

Commission on Country Life (1909) *Report of the Commission on Country Life.* United States Senate Document 705, 68th Congress, 2nd Session.

Commonwealth of Australia (1990) *Ecologically Sustainable Development: A Commonwealth Discussion Paper.* AGPS, Canberra.

—— (1992) *National Strategy for Ecologically Sustainable Development.* AGPS, Canberra.

—— (1996) *State of the Environment 1996.* Department of Environment, State and Territories, Canberra.

Commonwealth Scientific and Industrial Research Organisation (CSIRO) (2001) *Fine Resolution Assessment of Enhanced Greenhouse Climate Change in Victoria.* Atmospheric Division, CSIRO, Melbourne.

Conacher, A & Conacher, J (2000) *Environmental Planning and Management in Australia.* Oxford University Press, Melbourne.

Connell, RW (1987) *Gender and Power: Society, The Person and Sexual Politics.* Stanford University Press, Stanford.

—— (1995) *Masculinities.* Allen & Unwin, Sydney.

—— (2002) *Gender.* Blackwell Publishers, Malden MA; Polity Press, Cambridge, UK.

Coombes, M & Raybould, S (2001) Public policy and population distribution: Developing appropriate indicators of settlement patterns. *Environment and Planning C: Government and Policy* 19: 223–48.

Costanza, R et al. (1997) The value of the world's ecosystem services and natural capital. *Nature* 387: 253–60.

Council of Australian Governments (2002) *A National Action Plan for Salinity and Water Quality*. Agriculture, Fisheries and Forestry – Australia (AFFA) and Environment Australia (EA). At: <http://www.napswq. gov.au/publications/pubs/vital-resources.pdf>.

Courtney, P (2001) Potato price war a warning to big business. *Landline*. Australian Broadcasting Corporation. At: <http://www.abc.net.au/ landline/stories/s391826.htm>.

Cowlishaw, G (1988) *Black, White or Brindle: Race Relations in Rural Australia*. Cambridge University Press, Cambridge.

Cox, D & Foster, J (1990) Introduction. In D Cox (ed.) *Putting Rural Child Poverty on the Agenda*. Conference Proceedings, Dookie Agricultural College. VCOSS Papers No. 2. Victorian Council of Society Service, Collingwood, Victoria, pp. 1–4.

Cox, E & Caldwell, P (2000) Making policy social. In I Winter (ed.) *Social Capital and Public Policy in Australia*. Australian Institute of Family Studies, Melbourne, pp. 43–73.

Crawford, P & Crawford, I (2003) *Contested Country: A History of the Northcliffe Area, Western Australia*. University of Western Australia Press, Nedlands.

Crow, G & Allan, G (1994) *Community Life: An Introduction to Local Social Relations*. Harvester Wheatsheaf, London.

Cullen, P (2001) *Delivering Limnological Knowledge to the Water Industry*. SIL XXVIII Congress, Melbourne.

Curtis, A & Lockwood, M (1998) Natural resource policy for rural Australia. In J Prately & A Robertson (eds.) *Agriculture and the Environmental Imperative*. CSIRO, Collingwood.

Curtis, A, Robertson, A & Race, D (1998) Lessons from recent evaluations of natural resource management programs in Australia. *Australian Journal of Environmental Management* 5(2): 109–19.

Dahlstrom, M (1996) Young women in a male periphery: Experiences from the Scandinavian North. *Journal of Rural Studies* 12(3): 259–71.

Dairy Research and Development Corporation (DRDC) and National Land and Water Resources Audit (NLWRA) (2001) Dairy Catchments Australia. Unpublished document.

Dale, A & Hill, SB (2001) *At the Edge: Sustainable Development in the 21st Century*. UBC Press, Vancouver.

Daly, HE (1994) Operationalizing sustainable development by investing in natural capital. In AM Jansson, M Hammer, C Folke & R Constanza (eds.) *Investing in Natural Capital: The Ecological Economics Approach to Sustainability*. Island Press, Washington, DC, pp. 22–37.

—— (1995) On Wilfred Beckerman's critique of sustainable development. *Environmental Values* 4: 49–55.

Daly, HE & Cobb, JB (1989) *For the Common Good: Redirecting the Economy Toward Community, the Environment, and a Sustainable Future*. Beacon Press, Boston.

Daly, M (2000) The challenges for local government in the 21st century. In B Pritchard & P McManus (eds.), *Land of Discontent: The Dynamics of Change in Rural and Regional Australia*. UNSW Press, Sydney, pp. 195–217.

Dasgupta, S & Mitra, T (1983) Intergenerational equity and efficient allocation of exhaustible resources. *International Economic Review* 24: 133–53.

Davidson, AP (1991) Rethinking household livelihood strategies. In *Research in Rural Sociology and Development: Vol 5. A Research Study of Annual Household Strategies*, JAI Press, Greenwich and London, pp. 11–28.

Davison, G (2003a) Fatal attraction? The lure of technology and the decline of rural Australia, 1890–2000. *Tasmanian Historical Studies* 8(2): 234–52.

—— (2003b) The social survey and the puzzle of Australian sociology. *Australian Historical Studies* 34(121): 139–62

Dax, T (1996) Defining rural areas: International comparisons and the OECD indicators. *Rural Society* 6(3): 3–17.

Day, G (1998a) A community of communities? Similarity and difference in Welsh rural community studies. *The Economic and Social Review* 29(3): 233–57.

—— (1998b) Working with the grain? Towards sustainable rural and community development. *Journal of Rural Studies* 14(1): 89–105.

Dempsey, K (1990) *Smalltown: A Study of Social Inequality, Cohesion and Belonging.* Oxford University Press, Melbourne.

—— (1992) *A Man's Town: Inequality Between Women and Men in Rural Australia.* Oxford University Press, Melbourne.

Department of Family and Community Services (DFCS) (2000) *Participation Support for a More Equitable Society: The Interim Report of the Reference Group on Welfare Reform.* DFCS, Canberra.

Department of Foreign Affairs and Trade (DFAT) (1994) *Subsistence to Supermarket: Food and Agricultural Transformation in South-east Asia.* AGPS, Canberra.

Department of Health and Aged Care & the National Key Centre for Social Applications of Geographical Information Systems (GISCA) (1999) *Measuring Remoteness: Accessibility/Remoteness Index of Australia (ARIA).* Occasional Papers: New Series No. 6, Canberra.

Department of Infrastructure (1999) *Towns in Time: Population Change in Victoria's Towns and Rural Areas, 1981–96.* Department of Infrastructure, Melbourne.

Department of Primary Industries and Energy & Department of Human Services and Health (1994) *Rural, Remote and Metropolitan Areas Classification: 1991 Census Edition.* AGPS, Canberra.

Department of Primary Industries and Energy (DPIE) & Centre for Rural Social Research (CRSR) (1997) *Evaluation of the Rural Communities Access Program.* DPIE, Canberra.

Department of Transport and Regional Services (DOTRS) (2000) *Regional Solutions Programme.* DOTRS, Canberra. At: <http://www.regionalsolutions. gov.au>.

—— (2001) *Stronger Regions, a Stronger Australia.* DOTRS, Canberra.

——(2002) *Regional Solutions Programme: Fact Sheet 1.* At: <http://www. regionalsolutions.gov.au/facts_1.htm>.

——(2003) *Budget 2002–2003: Regional Highlights.* At: <http://www.dotrs. gov.au>.

Diamond, J (1997) *Guns, Germs and Steel: A Short History of Everybody for the Last 13 000 Years.* Vintage, London.

Dibden, J (2001) Regional Australia in transition. In J Dibden, M Fletcher & C Cocklin (eds.) *All Change!: Gippsland Perspectives on Regional Australia in Transition.* Monash Regional Australia Project, Monash University, Melbourne, pp. 1–17.

Dibden, J & Cocklin, C (2003) 'Tarra', Victoria. In C Cocklin & M Alston (eds.) *Community Sustainability in Rural Australia: A Question of Capital?* Centre for Rural Social Research, Charles Sturt University, Wagga Wagga, pp. 170–201.

Diesendorf, M & Hamilton, C (eds.) (1997a) *Human Ecology, Human Economy.* Allen & Unwin, Sydney.

—— (1997b) The ecologically sustainable development process in Australia. In M Diesendorf & C Hamilton (eds.) *Human Ecology, Human Economy: Ideas for an Ecologically Sustainable Future*. Allen & Unwin, Sydney, pp. 285–301.

Dimopoulos, M & Sheridan, M (2000) *Missed Opportunities: Unlocking the Future of Women in Agriculture*. Stage 2 Report. RIRDC, Canberra.

Dingle, A (1984) *Settling*. Fairfax, Syme & Weldon, McMahon's Point.

Dixit, A, Hammond, P & Hoel, M (1980) On Hartwick's rule for regular maximin paths of capital accumulation and resource depletion. *Review of Economic Studies* 47: 551–56.

Dixon, J (2002) *The Changing Chicken: Chooks, Cooks and Culinary Culture*. UNSW Press, Sydney.

Dobson, A (ed.) (1999) *Fairness and Futurity: Essays on Environmental Sustainability and Social Justice*. Oxford University Press, Oxford.

Dodson, L (2001) No wonder John Howard is a shade of green. *The Age*, 23 March, p. 15.

Doel, J (1999) Towards a supply-chain community? Insights from governance processes in the food industry. *Environment and Planning A* 31: 69–85.

Dollery, BE & Soul, S (2000) A note on Australian local government and regional economic and social inequalities. *Australian Journal of Social Issues* 35(2): 159–68.

Dominelli, L (2002) *Feminist Social Work Theory and Practice*. Palgrave, Basingstoke, Hampshire.

Dore, J & Woodhill J (1999) *Sustainable Regional Development Final Report: An Australia-wide Study of Regionalism Highlighting Efforts to Improve the Community, Economy and Environment*. Greening Australia, Canberra.

Dresner, A (2002) *The Principles of Sustainability*. Earthscan, London.

Durlauf, SN (2002) Bowling alone: A review essay. *Journal of Economic Behavior and Organization* 47: 259–73.

Eckersley, R (1998) *Measuring Progress: Is Life Getting Better?* CSIRO Publishing, Collingwood.

Eder, K (1996) *The Social Construction of Nature: A Sociology of Ecological Enlightenment*. Sage, London.

Edgar, D (2001) *The Patchwork Nation: Re-thinking Government – Re-building Community*. Harper Collins, Sydney.

Edwards, B (1998) Charting the discourse of community action: Perspectives from practice in rural Wales. *Journal of Rural Studies* 14(1): 63–77.

Eikeland, S (1999) New rural pluriactivity? Household strategies and rural renewal in Norway. *Sociologia Ruralis* 39(3): 359–76.

Elix, J & Lambert, J (1998) *Missed Opportunities: Harnessing the Potential of Women in Australian Agriculture*. RIRDC, DPIE, Canberra.

Environs Australia (1999) *Our Community Our Future: A Guide to Local Agenda 21*. Commonwealth of Australia, Canberra.

Epp, R (2001) The political de-skilling of rural communities. In R Epp & D Whitson, (eds.) *Writing Off the Rural West: Globalization, Governments and the Transformation of Rural Communities*. University of Alberta, Edmonton, pp. 301–24.

Evans, N, Morris, C & Winter, M (2002) Conceptualizing agriculture: A critique of post-productivism as the new orthodoxy. *Progress in Human Geography* 26(3): 313–32.

Eversole, R (2003) Value-adding community? Community economic development in theory and practice. *Rural Society* 13 (1): 72–86.

Fagan, R & Webber, M (1994) *Global Restructuring: The Australian Experience*. Oxford University Press, Melbourne.

Ferge, Z (2000) Poverty in Hungary and in Central and Eastern Europe. In D Gordon & P Townsend (eds.) *Breadline Europe: The Measurement of Poverty.* Policy Press, Bristol, pp. 267–306.

Fisher, A (1929) The Drift to the Towns. *Economic Record* 5: 234–52.

Fitzpatrick, B (1942) *The Case for Decentralisation and Defence.* Victorian Decentralisation League, Melbourne.

Flache, A & Macy, MW (1997) The weakness of strong ties: Collective action failure in a highly cohesive group. In P Doreian & FN Stokman (eds.) *Evolution of Social Networks.* Gordon & Breach, Amsterdam, pp. 19–44.

Fletcher, B (1969) Agriculture. In GJ Abbott & NB Nairn (eds.) *Economic Growth of Australia 1788–1821.* Melbourne University Press, Melbourne, pp. 191–218.

Forsyth, P (ed.) (1992) *Microeconomic Reform in Australia.* Allen & Unwin, Sydney.

Forsythe, WD (1942) *The Myth of the Open Spaces.* Melbourne University Press, Melbourne 1942.

Forth, G (2000) The future of Australia's declining country towns. *Regional Policy and Practice* 9(2): 4–10.

—— (2001) Following the yellow brick road and the future of Australia's declining country towns. In MF Rogers & YMJ Collins (eds.) *The Future of Australian Country Towns.* Centre for Sustainable Rural Communities, La Trobe University, Bendigo, pp. 72–82.

Forth, G & Howell, K (2002) Don't cry for me Upper Wombat: The realities of regional/small town decline in non-coastal Australia. *Sustaining Regions* 2(2): 4–11.

Foucault, M (1980) *Power/Knowledge: Selected Interviews and Other Writings 1972–1977.* Pantheon, New York.

—— (1983) The subject and power. In H Dreyfus & P Rabinow (eds.) *Michel Foucault: Beyond Structuralism and Hermeneutics.* University of Chicago Press, Chicago, pp. 208–20.

Friedmann, J (1992) *Empowerment: The Politics of Alternative Development.* Blackwell Publishers, Cambridge MA.

Frost, L (1991) *The New Urban Frontier: Urbanisation and City-Building in Australasia and the American West.* UNSW Press, Kensington.

Fulton, A & Clark, R (1996) Farmer decision making under contract farming in northern Tasmania. In D Burch, R Rickson & G Lawrence (eds.) *Globalization and Agri-food Restructuring: Perspectives from the Australasia Region.* Avebury, London, pp. 219–38.

Furuseth, O (1997) Sustainability issues in the industrialisation of hog production in the United States. In B Ilbery, Q Chiotti & T Rickard (eds.) *Agricultural Restructuring and Sustainability.* CAB International, Wallingford UK, pp. 293–312.

Fyfe, M (2002) Green issues out from under a cloud. *The Age,* 21 November, p. 11.

Gallie, WB (1956) Essentially contested concepts. *Proceedings of the Aristotelian Society* 56: 167–98.

Gammage, W (1986) *Narrandera Shire.* Narrandera Shire Council, Narrandera.

Garlick, S (1997) Regional economic development: New partnership challenges for local government. In B Dollery & N Marshall (eds.) *Australian Local Government: Reform and Renewal.* MacMillan Education Australia Pty Ltd, Melbourne, pp. 276–93.

Garnaut, J, Connell, P, Lindsay, R & Lindsay, V (2001) *Country Australia: Influences on Employment and Population Growth.* Australian Bureau of Agricultural and Resource Economics (ABARE), Canberra.

Garnaut, J & Lim-Applegate, H (1998) *People in Farming*. Australian Bureau of Agricultural and Resource Economics (ABARE), Canberra.

Garnaut, J, Rasheed, C & Rodrigues, C (1999) *Farmers at Work: The Gender Division*. Australian Bureau of Agricultural and Resource Economics (ABARE), Canberra.

Garnaut, R (1989) *Australia and the North East Asian Ascendancy: Report to the Prime Minister and the Minister for Foreign Affairs and Trade*. AGPS, Canberra.

Garnett, A & Lewis, P (1999) Trends in rural labour markets. Paper presented at the Country Matters national conference organised by the Bureau of Rural Sciences, Canberra, 20–21 May.

Garton, S (1996) *The Cost of War: Australians Return*. Oxford University Press, Melbourne.

Gerritson, R (2000) The management of government and its consequences for service delivery in regional Australia. In B Pritchard & P McManus (eds.) *Land of Discontent: The Dynamics of Change in Rural and Regional Australia*. UNSW Press, Sydney, pp. 123–39.

Gibson, K & Cameron, J (2001) Transforming communities: Towards a research agenda. *Urban Policy and Research* 19(1): 7–24.

Giddings, B, Hopwood, B & O'Brien, G (2002) Environment, economy and society: Fitting them together into sustainable development. *Sustainable Development* 10: 187–96.

Gippsland Development Ltd (2001) *Gippsland Economic Summit*. Gippsland Development Ltd, Sale, Victoria.

Giroux, H (2002) Global capitalism and the return of the garrison state. *Arena Journal* New Series, 19: 141–60.

Gleeson, B & Carmichael, C (2001) Responding to regional disadvantage: What can be learned from the overseas experience? Positioning paper prepared by the Australian Housing and Urban Research Institute, University of New South Wales and University of Western Sydney Research Centre, December.

Gleeson, T & Topp, V (1997) Broadacre farming today – forces for change. In Australian Bureau of Agricultural and Resource Economics (ABARE) *Outlook 97 Agriculture* ABARE, Canberra, pp. 53–74.

Glover, J, Harris, K & Tennant, S (1999) *A Social Health Atlas of Australia*. Second Edition. Public Health Information Development Unit, University of Adelaide, Open Book Publishers, Canberra.

Goodman, D & Redclift, M (1989) *The International Farm Crisis*. Macmillan Press, Basingstoke.

Goulburn Broken CMA (1999) *Draft Goulburn Broken Native Vegetation Management Strategy*. Goulburn Broken CMA, Shepparton.

Government of Western Australia (2000) *A Regional Development Strategy for Western Australia*. Government Printer, Perth.

—— (2002) *State Sustainability Strategy – Draft*. Government Printer, Perth.

Gow, J (1996) Structural adjustment in Australia revisited. *Rural Society* 6(1): 24–30.

Graham, BD (1966) *The Formation of the Australian Country Parties*. Australian National University Press, Canberra.

Granovetter, M (1972) The strength of weak ties. *American Journal of Sociology* 78(6): 1360–80.

Gray, A (1997) *Growth of the Aboriginal and Torres Strait Islander Population, 1991–2001 and Beyond*. CAEPR Discussion Paper No. 150/1997. The Australian National University, Canberra.

Gray, I (1991) *Politics in Place: Social Power Relations in an Australian Country Town*. Cambridge University Press, Cambridge.

—— (1996) The detraditionalization of farming. In D Burch, R Rickson, & G Lawrence (eds.) *Globalization and Agri-food Restructuring: Perspectives from the Australasia Region*. Avebury, London, pp. 91–106.

Gray, I & Lawrence, G (2001a) *A Future for Regional Australia: Escaping Global Misfortune*. Cambridge University Press, Melbourne.

—— (2001b) Neoliberalism, individualism and prospects for regional renewal. *Rural Society* 11(3): 283–97.

Gray, I, Lawrence, G & Dunn, T (1993) *Coping with Change: Australian Farmers in the 1990s*. Centre for Rural Social Research, Charles Sturt University, Wagga Wagga.

Gray, I, Phillips, E & Dunn, A (2000). Aspects of rural culture and the use of conservation farming. In AD Shulman & RJ Price (eds.) *Case Studies in Increasing the Adoption of Sustainable Resource Management Practice*. Land and Water Resources Research and Development Corporation, Canberra, pp. 34–47.

Gray, I, Williams, R & Phillips, E (2002) Community capacity for agricultural sustainability: Issues for policy amid tensions between community and leadership. Paper given at the International Sociological Association conference, Brisbane.

Haberkorn, G, Kelson, S, Tottenham, R & Magpantay, C, (2003) *Country Matters: Social Atlas of Rural and Regional Australia*. AGPS, Canberra.

Hain, M, Cocklin, C & Gibbs, D (2002) Regulating biosciences: The Gene Technology Act 2000. *Environmental and Planning Law Journal* 19(3): 163–79.

Hall, AR (1968) *The Stock Exchange of Melbourne and the Victorian Economy, 1852–1900*. ANU Press, Canberra.

Hall, R, Thorns, D & Willmott, WE (1984) Community, class and kinship: Bases for collective action within localities. *Environment and Planning D: Society and Space* 2(2): 201–15.

Halliday, J & Little, J (2001) Amongst women: Exploring the reality of rural childcare. *Sociologia Ruralis* 41(4): 423–37.

Halligan, J & Paris, C (1984) The politics of local government. In J Halligan & C Paris (eds.) *Australian Urban Politics*. Longman Cheshire, Melbourne, pp. 58–72.

Hamer, D (1990) *New Towns in the New World: Images and Perceptions of the Nineteenth Century Urban Frontier*. Columbia University Press, New York.

Harris, HL (1948) The implications of decentralisation. In HL Harris et al. (eds.) *Decentralisation: Papers Read at the 1949 Summer School of Australian Institute of Political Science*. Angus & Robertson, Sydney, pp. 1–23.

Harrison, H (1997) *Trends in the Delivery of Rural Health, Education and Banking Services*. National Farmers' Federation, Kingston, ACT.

Hart, M (2000) Key term: Community capital. *Sustainable Measures*. At: <http://www.sustainablemeasure.com/Sustainability/KeyTermCommC apital.html>.

Hartwick, JM (1977) Intergenerational equity and the investing of rents from exhaustible resources. *American Economic Review* 67: 972–74.

—— (1978a) Investing returns from depleting renewable resource stocks and intergenerational equity. *Economic Letters* 1: 85–88.

—— (1978b) Substitution among exhaustible resources and intergenerational equity. *Review of Economic Studies* 45: 347–54.

—— (1990) Natural resources, national accounting and economic depreciation. *Journal of Public Economics* 43: 291–304.

—— (1995) Consumption paths in open economies with exhaustible resources. *Review of International Economics* 3: 275–83.

Harvey, D (1989) From managerialism to entrepreneurialism: The transformation in urban governance in late capitalism. *Geografiska Annaler* 71b(1): 3–17.

Haslam-McKenzie, F (1999) *The Impact of Declining Infrastructure in Rural Western Australia.* Rural Industries Research and Development Corporation, Canberra.

—— (2003a) Strategies for getting women back on centre stage. In R O'Hagan, M Alston & S Spriggs (eds.) *Setting the Agenda for Rural Women: Research Directions. Conference Proceedings and Recommendations.* Centre for Rural Social Research, Charles Sturt University, Wagga Wagga, pp. 36–46.

—— (2003b) A review of the status of women in rural, regional and remote Australia: Identifying leadership strategy successes and challenges. In R O'Hagan, M Alston & S Spriggs (eds.) *Setting the Agenda for Rural Women: Research Directions. Conference Proceedings and Recommendations.* Centre for Rural Social Research, Charles Sturt University, Wagga Wagga, pp. 133–49.

—— (2003c) Economic restructuring and regional development in the central wheatbelt of Western Australia. Unpublished PhD thesis, University of Western Australia, Perth.

Haughton, G (1999) Community economic development: Challenges of theory, method and practice. In G Haughton (ed.) *Community Economic Development.* The Stationary Office, London, pp. 3–22.

Hawe, P & Shiell, A (2000) Social capital and health promotion: A review. *Social Science and Medicine* 51: 871–85.

Heelas, P, Lash, S & Morris, P (eds.) (1996) *Detraditionalisation: Critical Reflections on Authority and Identity.* Blackwell, Massachusetts.

Herbert-Cheshire, L (2000) Contemporary strategies for rural community development in Australia: A governmentality perspective. *Journal of Rural Studies* 16(2): 203–15.

—— (2001) 'Changing people to change things': Building capacity for self-help in natural resource management – a governmentality perspective. In G Lawrence, V Higgins & S Lockie (eds.) *Environment, Society and Natural Resource Management: Theoretical Perspectives from Australasia and the Americas.* Edward Elgar, Cheltenham, pp. 270–82.

Herbert-Cheshire, L & Lawrence, G (2003) Monto, Queensland. In C Cocklin & M Alston (eds.) *Community Sustainability in Rural Australia: A Question of Capital?* Charles Sturt University, Centre for Rural Social Research, Wagga Wagga, pp. 10–37.

Higgins, V (1998) Breaking down the divisions? A case study of political power and changing rural local government representation. *Rural Society* 7(3/4): 51–58.

Higgins, V & Lockie, S (2001a) Getting big and getting out. Government policy, self-reliance and farm adjustment. In S Lockie & L Bourke (eds.) *Rurality Bites: The Social and Environmental Transformation of Rural Australia.* Pluto Press, Sydney, pp. 178–90.

—— (2001b) Neo-liberalism and the governing of natural resource management in Australia. In J Dibden, M Fletcher & C Cocklin (eds.) *All Change! Gippsland Perspectives on Regional Australia in Transition.* Monash Regional Australia Project, Monash University, Melbourne, pp. 97–105.

Hillery, GA (1955) Definitions of community: Areas of agreement. *Rural Sociology* 20(2): 111–23.

Hindmarsh, R, & Lawrence, G (eds.) (2001) *Altered Genes II: The Future?* Scribe, Melbourne.

Hindmarsh, R, Lawrence, G & Norton, J (eds.) (1998) *Altered Genes – Reconstructing Nature: The Debate.* Allen & Unwin, Sydney.

Hirst, J (1967) Centralization reconsidered: The South Australian Education Act of 1875. *Historical Studies* 13(40): 42–59.

Hobcraft, J (2001) *The Roles of Schooling and Educational Qualifications in the Emergence of Adult Social Exclusion.* STICERD Publications, London.

Hoggart, K & Buller, H (1987) *Rural Development: A Geographical Perspective.* Croom Helm, London.

Holland, A (1997) Substitutability: Or why strong sustainability is weak and absurdly strong sustainability is not absurd. In J Foster (ed.) *Valuing Nature: Ethics, Economics and the Environment.* Routledge, London, pp. 119–34.

Hollander, G (2004) Agricultural trade liberalization, multifunctionality, and sugar in the south Florida landscape *Geoforum* (in press).

Holmes, J (1987) Population. In D Jeans (ed.) *Space and Society.* Sydney University Press, Sydney, pp. 24–48.

—— (2002) Diversity and change in Australia's rangelands: A post-productivist transition with a difference? *Transactions of the Institute of British Geographers* 27(3): 362–84.

—— (2004) Personal communication.

Holmes, JH (1994) Coast versus inland: Two different Queenslands. *Australian Geographical Studies* 32(2): 17–82.

Holt, J (1946) *Wheat Farms of Victoria: A Sociological Survey.* Melbourne University Press, Melbourne.

Hooper, S, Martin, P, Love, G & Fisher, B (2002) 'Get big or get out': Is this mantra still appropriate for the new century? Paper given at the 24th Biennial Conference of the Australian Society of Animal Production, Adelaide, 11 July. At: <http://abareonlineshop.com/product.asp?prodid=12385>.

Hopkins, P (2001) Plantations in full flourish. *The Age*, 12 March, Business, p. 5.

Hudson, P (1989) Change and adaptation in four rural communities in New England, NSW. *Australian Geographer* 20(1): 54–64.

Hugo, G (2001) Defining social catchments in Australia: Some preliminary ideas. PowerPoint slide presentation for the Serving Regional Australia Project, DTRS and BRS.

—— (2003) What is Happening to Rural and Regional Populations? In *First National Conference on the Future of Australia's Country Towns, August 2000.* The Regional Institute Ltd. At: <http://www.regional.org.au/au/countrytowns/keynote/hugo.htm>.

Hugo, GJ (1971) Internal migration in South Australia, 1961–66. Unpublished MA Thesis. Flinders University of South Australia, Adelaide.

—— (1976) Demographic and social implications of Australian policy on population distribution and redistribution. In RJ Pryor (ed.) *Population Redistribution: Policy Research.* Studies in Migration and Urbanisation, No. 2 Australian National University, Canberra, pp. 66–78.

—— (2001) A century of population change in Australia. In Australian Bureau of Statistics (ABS) *2002 Year Book Australia*, No. 83, Catalogue No. 1301.0. Australian Bureau of Statistics, Canberra, pp. 169–210.

—— (2003) Recent trends in internal migration and population redistribution in Australia. Paper prepared for presentation at the Population Association of America 2003 Annual Meeting, Minneapolis, 1–3 May.

Hugo, GJ & Bell, M (1998) The hypothesis of welfare-led migration to rural areas: The Australian case. In P Boyle & K Halfacree (eds.) *Migration into Rural Areas-Theories and Issues*. John Wiley & Sons, West Sussex, pp. 107–33.

Hugo, GJ, Griffith, D, Rees, P, Smailes, P, Badcock, B & Stimson, R (1997) *Rethinking the ASGC: Some Conceptual and Practical Issues*. Final Report for Review of the Australian Standard Geographical Classification Project. Monograph Series 3/1997, National Key Centre for Social Applications of Geographical Information Systems, Adelaide.

Hugo, GJ & Smailes, PJ (1985) Urban-rural migration in Australia: A process view of the turnaround. *Journal of Rural Studies* 1(1): 10–11.

Human Rights and Equal Opportunity Commission (HREOC) (1999) *Bush Talks*. HREOC, Sydney.

—— (2000a) *'Emerging Themes': National Inquiry into Rural and Remote Education*. HREOC, Sydney.

—— (2000b) *Education Access: National Inquiry into Rural and Remote Education*. Human Rights and Equal Opportunities Commission, Sydney.

Ife, J (1995) *Community Development: Creating Community Alternatives – Vision, Analysis and Practice*. Longman Australia, Melbourne.

—— (2001) *Human Rights and Social Work: Towards Rights-Based Practice*. Cambridge University Press, Cambridge and Melbourne.

Ilbery, BW & Bowler, I (1998) From agricultural productivism to post-productivism. In B Ilbery (ed.) *The Geography of Rural Change*. Longman, London, pp. 57–84.

Institute of Land and Food Resources (2000) *Socioeconomic Impact of Changing Land Use in South-West Victoria*. University of Melbourne, Melbourne.

Jessop, B (1995) The regulation approach, governance and post-Fordism: Alternative perspectives on economic and political change. *Economy and Society* 24: 307–33.

Jones, A (1991) Farmers and their towns. In AF Denholm, S Marsden & K Round, (eds.) *Terowie Workshop: Exploring the History of South Australian Country Towns*. Department of History, University of Adelaide, Adelaide, pp. 94–111.

Jones, A, Reddel, T & Smyth, P (2001) Local governance and social inclusion: Towards an effective place policy framework. In T Reddel (ed.) *Local Governance and Social Inclusion: Places and Policies*. School of Social Work and Social Policy Occasional Paper No. 3. The University of Queensland, Brisbane, pp. 57–83.

Jones, O (1995) Lay discourses of the rural: Developments and implications for rural studies. *Journal of Rural Studies* 11(1): 35–49.

Kahn, A (1966) The tyranny of small decisions: Market failures, imperfections and the limits of economics. *Kyklos* 19: 23–47.

Keane, M (1990) Economic development capacity amongst small rural communities. *Journal of Rural Studies* 6(3): 291–301.

Kelly, P (1992) *The End of Certainty: The Story of the 1980s*. Allen & Unwin, Sydney.

Kenyon, P & Black, A (eds.) (2001) *Small Town Revival: Overview and Case Studies: A Report for the Rural Industries Research and Development Corporation*. RIRDC, Canberra.

Knutson, R, Romain, R, Anderson, D & Richardson, J (1997) Farm-level consequences of Canadian and US dairy policies. *American Journal of Agricultural Economics* 79: 1563–72.

Krugman, P (1991) *Geography and Trade*. Leuven University Press and MIT, Leuven and Cambridge, MA.

Labonte, R (1999) Social capital and community development: Practitioner emptor. *Australian and New Zealand Journal of Public Health* 23(4): 430–34.

Lake, M (1987) *The Limits of Hope: Soldier Settlement in Victoria, 1915–38.* Oxford University Press, Melbourne.

Langton, M (2002). A new deal? Indigenous development and the politics of recovery. Dr Charles Perkins Memorial Oration, University of Sydney, 4 October. At: <http://www.media.usyd.edu.au/speeches/2002/langton.pdf>.

Lannin, D (1997) An evaluation of the Foundations for Leadership course. Unpublished Bachelor of Agribusiness honours thesis, Muresk Institute of Agriculture, Curtin University of Technology, Northam.

Laoire, C (2001) A matter of life and death? Men, masculinities and staying 'behind' in rural Ireland. *Sociologia Ruralis* 41(2): 220–36.

Lash, S & Urry, J (1994) *Economies of Signs and Space.* Sage, London.

Laughton, AM (1916) Evidence of Alexander M. Laughton to Victorian Legislative Assembly Select Committee on the Drift of Population from the Country Districts to the City, 12 September 1916. Copy of typescript evidence supplied by Parliamentary Papers Office, Parliament House, Melbourne.

Lawless, P (2001) Community economic development in urban and regional regeneration: Unfolding potential or justifiable scepticism? *Environment and Planning C: Government and Policy* 19: 135–55.

Lawrence, G (1987) *Capitalism and the Countryside: The Rural Crisis in Australia.* Pluto Press, Sydney.

—— (1995) *Futures for Rural Australia: From Agricultural Productivism to Community Sustainability.* Centre for Rural Economic and Social Research, Central Queensland University, Rockhampton.

—— (2003) Sustainable regional development: Recovering lost ground. Paper presented at the 'Social Dimensions of the Triple Bottom Line in Rural Australia' one day seminar, Bureau of Rural Sciences, Canberra, 26 February.

Lawrence, G, Gray, I & Stehlik, D (1997) Changing spaces: The effects of macro-social forces on rural Australia. Paper presented at the XVII Congress of the European Society for Rural Sociology, Mediterranean Agronomic Institute at Chania, Crete, 25–29 August.

Lawrence, G, Lyons, K & Lockie, S (1999) 'Healthy for you, healthy for the environment': Corporate capital, farming practice and the construction of 'green' foods. *Rural Society* 9(3): 543–54.

Lawrence, G, Vanclay, F & Furze, B (eds.) (1992) *Agriculture, Environment and Society: Contemporary Issues for Australia.* Macmillan, Melbourne.

Leader (1918). 21 June.

Liepins, R (2000a) New energies for an old idea: Reworking approaches to 'community' in contemporary rural studies. *Journal of Rural Studies* 16(1): 23–35.

—— (2000b) Exploring rurality through 'community': Discourses, practices and spaces shaping Australian and New Zealand rural 'communities'. *Journal of Rural Studies* 16(3): 325–41.

—— (2000c) Making men: The construction and representation of agriculture-based masculinities in Australia and New Zealand. *Rural Sociology* 65(4): 605–19.

Levitas, R (2000). What is social exclusion? In D Gordon & P Townsend (eds.) *Breadline Europe: The Measurement of Poverty.* Policy Press, Bristol, pp. 357–84.

Linn, R (1999) *Battling the Land: 200 Years of Rural Australia.* Allen & Unwin, Sydney.

Little, J (1994) Gender relations and the rural labour process. In S Whatmore, T Marsden & P Lowe (eds.) *Gender and Rurality.* David Fulton Publishers Ltd, London, pp. 11–30.

Little, P & Watts, M (eds.) (1994) *Living Under Contract: Contract Farming and Agrarian Transformation in Sub-Saharan Africa.* University of Wisconsin Press, Madison.

Lloyd, CJ & Troy PN (eds.) (1981) *Innovation and Reaction: The Life and Death of the Federal Department of Urban and Regional Development.* Allen & Unwin, Sydney.

Lobao, L (1996) A sociology of the periphery versus a peripheral sociology: Rural sociology and the dimension of space. *Rural Sociology* 61(1): 77–102.

Lockie, S (2000) Crisis and conflict: Shifting discourses of rural and regional Australia. In B Pritchard & P McManus (eds.) *Land of Discontent: The Dynamics of Change in Rural and Regional Australia.* UNSW Press, Sydney, pp. 14–32.

—— (2001a) Positive futures for rural Australia. In S Lockie & L Bourke (eds.) *Rurality Bites: The Social and Environmental Transformation of Rural Australia.* Pluto Press, Sydney, pp. 287–99.

—— (2001b) Agriculture and environment. In S Lockie & L Bourke (eds.) *Rurality Bites: The Social and Environmental Transformation of Rural Australia.* Pluto Press, Sydney, pp. 229–42.

Lockie, S & Bourke, L (eds.) (2001) *Rurality Bites: The Social and Environmental Transformation of Rural Australia.* Pluto Press, Sydney.

Lockie, S, Dale, A, Taylor, B & Lawrence, G (2000a) Capacity for change: Testing a model for the inclusion of social indicators in Australia's national land and water resources audit. Unpublished manuscript.

Lockie, S, Lyons, K & Lawrence, G (2000b) Constructing 'green' foods: Corporate capital, risk and organic farming in Australia and New Zealand. *Agriculture and Human Values* 17: 315–22.

Lockie, S & Vanclay, F (eds.) (1997) *Critical Landcare.* Centre for Rural Social Research, Charles Sturt University, Wagga Wagga.

Lothian, A (2002) Australian attitudes towards the environment: 1991–2001. *Australian Journal of Environmental Management* 9(1): 45–61.

Lowe, P, Murdoch, J, Marsden, T, Munton, R & Flynn, A (1993) Regulating the new rural spaces: The uneven development of land. *Journal of Rural Studies* 9(3): 205–22.

Lusthaus, C, Adrien, MH & Perstinger, M (1999) *Capacity Development: Definitions, Issues and Implications for Planning, Monitoring and Evaluation.* Universalia Occasional Paper 35. Montreal. At: <http://www.universalia.com/files/occas35.pdf>.

Lynn, M (2001) The community development capacity of corporatised rural human service organisations. Paper delivered to the RSWAG 'Building Bridges' conference, Beechworth Campus, La Trobe University.

Lyons, K (1998) Understanding organic farm practice: Contributions from eco-feminism. In D Burch, G Lawrence, R Rickson, & J Goss (eds.) *Australasian Food and Farming in a Globalised Economy: Recent Developments and Future Prospects.* Department of Geography and Environmental Science, Monash University, Melbourne, pp. 57–68.

McCarty, JW (1964) The staple approach to Australian economic history. *Business Archives and History* 4: 1–24.

—— (1978) Australian capital cities in the nineteenth century. In JW McCarty & CB Schedvin (eds.) *Australian Capital Cities: Historical Essays.* Sydney University Press, Sydney, pp. 9–25.

McEwen, E (n.d.) Country towns. Unpublished paper for the 1888 Bicentennial History Project.

Macgarvey, A & O'Toole, K (2003) The mill and the mount: Rural women in community economic development. In R O'Hagan, M Alston & S Spriggs (eds.) *Setting the Agenda for Rural Women: Research Directions.* Conference Proceedings and Recommendations. Centre for Rural Social Research, Charles Sturt University, Wagga Wagga, pp. 186–97.

McIntyre, AJ & McIntyre, JJ (1944) *Country Towns of Victoria: A Social Survey.* Melbourne University Press with Oxford University Press, Melbourne.

Macintyre, S (1985) *Winners and Losers.* Allen & Unwin, Sydney.

McKenzie, D (1997) Size counts as 15 000 small farms vanish. *The Australian.* 25 June, p. 3.

McKenzie, F (1994) Population decline in non-metropolitan Australia. *Urban Policy and Research* 12(4): 253–63.

Macklin, M (1995) Breaching the idyll: Ideology, intimacy and social service provision in a rural community. In P Share (ed.) *Communication and Culture in Rural Areas.* Centre for Rural Social Research, Charles Sturt University, Wagga Wagga, pp. 71–86.

Maclaren, J (1995) Environmental effects of plantations – a review. In D Hammond (ed.) *Forestry Handbook.* New Zealand Institute of Forestry, Christchurch.

—— (1996) *Environmental Effects of Planted Forests.* FRI Bulletin 198, N.Z. Forest Research Institute, Rotorua.

McMichael, P (ed.) (1994) *The Global Restructuring of Agro-food Systems.* Cornell University Press, Ithaca.

—— (1996) Globalization: Myths and realities. *Rural Sociology* 61(1): 25–55.

—— (2000) *Development and Social Change: A Global Perspective.* Second Edition. Pine Forge, Thousand Oaks.

McMichael, P & Lawrence, G (2001) Globalising agriculture: Structures of constraint for Australian farming. In S Lockie & L Bourke (eds.) *Rurality Bites: The Social and Environmental Transformation of Rural Australia.* Pluto Press, Sydney, pp. 153–64.

McNay, L (2000). *Gender and Agency: Reconfiguring the Subject in Feminist and Social Theory.* Polity Press, Cambridge.

Macquarie, L (1956) *Journals of his Tours of New South Wales and Van Diemen's Land, 1810–1822.* Trustees of the Public Library of New South Wales, Sydney.

Mansfield, R (1841) *Analytical View of the Census of New South Wales for the Year 1841.* Kemp & Fairfax, Sydney.

Marsden, T (2003) *The Condition of Rural Sustainability.* Royal Van Gorcum, Assen (Netherlands).

Marsden, T, Murdoch, J, Lowe, P, Munton, R & Flynn, A (1993) *Constructing the Countryside.* UCL Press, London.

Massey, D (1994) *Space, Place and Gender.* Polity Press, Cambridge.

Martin, P, Hooper, S, Blias, A, Hanna, N & Ford, M (2003) Farm financial performance: Drought in 2002–03 cut incomes after an excellent year in 2001–02. In Australian Bureau of Agricultural and Resource Economics (ABARE) *Australian Farm Surveys Report* 2003. ABARE, Canberra.

Martin, P & Woodhill, J (1995) Landcare in the balance: Government roles and policy issues in sustaining rural environments. *Australian Journal of Environmental Management* 2: 173–83.

Martindale, D & Hanson, RG (1969) *Small Town and the Nation: The Conflict of Local and Translocal Forces.* Greenwood Publishing, Westport.

Maude, A (2002) An area based strategy for social inclusion in regional cities: An assessment. Paper prepared for the AHURI Southern Research Centre Seminar on Social Inclusion and Housing: Developing Research and Policy Agendas, Adelaide, 20 June.

Meadows, DH, Meadows, DL & Randers, J (1992) *Beyond the Limits: Global Collapse or a Sustainable Future.* Earthscan, London.

Meinig, DW (1970) *On the Margins of the Good Earth: The South Australian Wheat Frontier 1869–1884.* Rigby, Adelaide. (First published Rand McNally, Chicago 1962.)

Merrett, D (1977) Australian capital cities in the twentieth century. *Monash Papers in Economic History* 4: 171–98.

—— (1978) Australian capital cities in the twentieth century. In McCarty, JW & Schedvin, CB (eds.) *Australian Capital Cities: Historical Essays.* Sydney University Press, Sydney, pp. 171–98.

Metcalfe, A (1988) *For Freedom and Dignity.* Allen & Unwin, Sydney.

Miller, C (2001) Liberals set up own green group. *The Age,* 1 April, p. 4.

Miller, L (1994) Agribusiness, contract farmers and land use sustainability in North West Tasmania. Paper presented at the Agri-food Network Workshop, University of Sydney, 12–13 November.

—— (1996) Contract farming under globally-oriented and locally-emergent agribusiness in Tasmania. In D Burch, R Rickson, & G Lawrence (eds.) *Globalization and Agri-food Restructuring: Perspectives from the Australasia Region.* Avebury, London, pp. 203–18.

Ministry of Agriculture and Fisheries (MAF) (1994) *Forestry and Community: A Scoping Study of the Impact of Exotic Forestry on Rural New Zealand Communities Since 1980.* MAF Technical Paper 94/8, MAF Policy, Wellington.

Mobbs, C & Moore, K (2002) Foreward. In Land and Water Australia *Property: Rights and Responsibilities – Current Australian Thinking.* Land and Water Australia, Canberra, p. v.

Molotch, H (1976) The city as a growth machine. *American Journal of Sociology* 82(2): 309–30.

Monash Regional Australia Project (MRAP) (2001) *Social Capability in Rural Victoria: The Food and Agriculture and Natural Resource Management Sectors.* Department of Natural Resources and Environment, Bendigo.

Monk, A (1998) The Australian organic basket and the global supermarket. In D Burch, G Lawrence, R Rickson & J Goss (eds.) *Australasian Food and Farming in a Globalised Economy: Recent Developments and Future Prospects.* Department of Geography and Environmental Science, Monash University, Melbourne, pp. 69–80.

Moreira, M (1998) The dynamics of global capital and its consequences for agriculture and rural spaces. Paper presented to the Sociology of Agriculture and Food section, XIVth World Congress of Sociology, Montreal, 26 July–1 August.

Morvaridi, B (1998) Trade, contract farming and diversification: Environmental problems in a sugar-beet growing region of Turkey. Paper presented to the Sociology of Agriculture and Food section, XIVth World Congress of Sociology, Montreal, 26 July–1 August.

Mowbray, M (2000) Community development and local government: An Australian response to globalization and economic fundamentalism. *Community Development Journal* 35(3), 215–23.

Munn, P & Munn, T (2003) Rural social work: Moving forward. *Rural Society* 13(1): 22–34.

Murdoch, J (1997) The shifting territory of government: Some insights from the Rural White Paper. *Area* 29(2): 109–18.

Murphy, P (2002) Sea change: Reinventing rural and regional Australia. *Transformations* 2. At: <http://www.ahs.cqu.edu.au/transformations/journal/pdf/no2/murphy.pdf>.

Nalson, JS & Craig, RA (1989) Rural Australia. In S Encel & M Berry (eds.) *Selected Readings in Australian Society.* Longman Cheshire, Melbourne, pp. 311–43.

Napier, R (1997) Business structures for the future. In Australian Bureau of Agricultural and Resource Economics (ABARE) *Outlook 97, Volume 2 Agriculture.* ABARE, Proceedings of the National Agricultural and Resources Outlook Conference, Canberra, pp. 83–92.

National Farmers' Federation (NFF) & Australian Conservation Foundation (ACF) (2000) *A National Scenario for Strategic Investment.* At: <http://www.nff.org.au/pages/pub/5point.htm>

National Institute of Economic and Industry Research (NIEIR) (1998) *State of the Regions: A Report to the Australian Local Government Association 1998 Regional Cooperation and Development Forum.* National Institute of Economic and Industry Research, Melbourne.

National Land and Water Resources Audit (NLWRA) (2001a) *Australian Water Resources Assessment 2000.* Commonwealth of Australia, Canberra.

—— (2001b) *Australian Dryland Salinity Assessment 2000.* Commonwealth of Australia, Canberra.

National Natural Resource Management Policy Statement (NNRMPS) (1999) *Managing Natural Resources in Rural Australia for a Sustainable Future: A Discussion Paper for Developing a National Policy.* At: <http://www.napswq.gov.au/publications/nrm-discussion.html>.

National Rural Health Alliance (1998) *Blueprint for Rural Development: Discussion Paper.* National Rural Health Alliance, Deakin West, ACT.

Neumayer, E (1999) *Weak versus Strong Sustainability: Exploring the Limits of Two Opposing Paradigms.* Edward Elgar, Cheltenham.

Newnham, L & Winston, G (1997) The role of councillors in a changing local government arena. In B Dollery & N Marshall (eds.) *Australian Local Government: Reform and Renewal.* Macmillan, Melbourne, pp. 105–24.

New South Wales Premier's Department Strengthening Communities Unit (2001) *Strengthening Rural Communities Manual.* Strengthening Communities Unit, New South Wales Premier's Department, Sydney.

Oakley, A (2002) *Gender on Planet Earth.* Polity Press, Cambridge.

O'Connor, KB & Stimson, RJ (1996) Convergence and divergence of demographic and economic trends. In PW Newton & M Bell (eds.) *Population Shift: Mobility and Change in Australia.* AGPS, Canberra, pp. 108–25.

Odum, W (1982) Environmental degradation and the tyranny of small decisions *BioScience* 2(9): 728–29.

O'Hara, P (1994) Constructing the future: Cooperation and resistance among farm women in Ireland. In S Whatmore, T Marsden & P Lowe (eds.) *Gender and Rurality.* David Fulton Publishers, London, pp. 50–68.

O'Neill, J, Turner, RK & Bateman, IJ (eds.) (2001) *Environmental Ethics and Philosophy: Managing the Environment for Sustainable Development.* Edward Elgar, Cheltenham.

O'Loughlin, T (2004) US Help on Kyoto Alternative. *Australian Financial Review,* 16 January, p. 10.

Oman, C (1996) *The Policy Challenges of Globalisation and Regionalisation.* OECD Development Centre, OECD, Paris.

Oxley, HG (1978) *Mateship in Local Organization*. Second edition. University of Queensland Press, Brisbane, pp. 105–24.

Pahl, RE (1966) The rural-urban continuum. *Sociologia Ruralis* 6: 299–327.

Panelli, R (2001) Narratives of community and change in a contemporary rural setting: The case of Duaringa, Queensland. *Australian Geographical Studies* 39(2): 156–66.

Pastor, Manuel (2001) *Building Social Capital to Protect Natural Capital: The Quest for Environmental Justice*. Working Paper Series, Number 11, Program on Development, Peacebuilding, and the Environment Political Economy Research Institute (PERI). University of Massachusetts, Amherst.

Peatling, S & Riley, M (2004) Greenhouse Gas Scheme Gets the Axe. *Sydney Morning Herald*, 12 January, p. 1.

People Together Project (2000) *The Power of Community: Celebrating and Promoting Community in Victoria*. Local Governance Association, Melbourne.

Pierce, J (1992) Progress and the biosphere: The dialectics of sustainable development. *The Canadian Geographer* 36: 306–19.

Porter, ME (1998) Location, clusters, and the 'new' microeconomics of competition. *Business Economics* 33(1): 7–13.

Portes, A (1998) Social capital: Its origins and applications in modern sociology. *Annual Review of Sociology* 24: 1–24.

Potter, C & Burney, J (2002) Agricultural multifunctionality in the WTO – legitimate non-trade concern or disguised protectionism? *Journal of Rural Studies* 18: 35–47.

Powell, F (2001) *The Politics of Social Work*. SAGE, London.

Powell, J (ed.) (1974) *The Making of Rural Australia – Environment, Society and Economy: Geographical Readings*. Sorrett Publishing, Melbourne.

—— (1988) *An Historical Geography of Modern Australia: The Restive Fringe*. Cambridge University Press, Melbourne.

—— (1993) *Griffith Taylor and 'Australia Unlimited'*. University of Queensland Press, St Lucia.

Pretty, J (1999) Current challenges for agricultural development. Paper presented to the Kentucky Cooperative Extension Service Conference, Lexington, Kentucky, January 13–15, 1999. At: <http://www.uky.edu/Agriculture/AgPrograms/australia/pretty_sustain_ag/prettyall-p.html>.

—— (2002) People, livelihoods and collective action in biodiversity management. In T O'Riordan & S Stoll-Kleeman (eds.) *Biodiversity, Sustainability and Human Communities: Protecting Beyond the Protected*. Cambridge University Press, Cambridge.

Pritchard, B (1999) Australia as the supermarket to Asia? Governments, territory, and political economy in the Australian agri-food system. *Rural Sociology* 64(2): 284–301.

Pritchard, B & Burch, D (2003) *Agrifood Globalization in Perspective: International Restructuring in the Processing Tomato Industry*. Ashgate, Aldershot.

Pritchard, B & McManus, P (eds.) (2000) *Land of Discontent: The Dynamics of Change in Rural and Regional Australia*. UNSW Press, Sydney.

Productivity Commission (1999) *Impact of Competition Policy Reforms on Rural and Regional Australia*. Commonwealth of Australia, Canberra.

—— (2003a) *Social Capital: Reviewing the Concept and Its Policy Implications*. Commonwealth of Australia, Canberra.

—— (2003b) *Impacts of Native Vegetation and Biodiversity Regulations* (Draft Report). Ausinfo, Canberra.

Putnam, R (1993) The prosperous community: Social capital and public life. *The American Prospect* Spring: 35–42.

Ramp, W & Koc, M (2001) Global investment and local politics: The case of Lethbridge. In R Epp & D Whitson (eds.) *Writing Off the Rural West: Globalization, Governments and the Transformation of Rural Communities.* University of Alberta, Edmonton, pp. 53–70.

Rapport, D, Costanza, R & McMichael, A (1998) Assessing ecosystem health. *TREE* 13(10): 397–402.

Ray, C (2000) The EU LEADER programme: Rural development laboratory. *Sociologia Ruralis* 40(2): 163–71.

Rees, P & Fischer, T (2002) *Tim Fischer's Outback Heroes and Communities that Count.* Allen & Unwin, Sydney.

Reeve, I & Black, A (1993) *Australian Farmers' Attitudes to Rural Environmental Issues.* Rural Development Centre, University of New England, Armidale.

Reeves, William Pember (1969) *State Experiments in Australia and New Zealand.* Volume 1. Macmillan, Melbourne. (First published Grant Richards, London 1902.)

Reimer, W & Aipedale, P (2000) *The New Rural Economy in Canada.* Western Agri-Food Institute Internet Colloquium 2000 (Canada). At: <ftp://132.205.87.156/western_agrifood_institute/Rural_Restructurin g_canada.pdf>

Richardson, NH (1994) Making our communities sustainable: The central issue is will. *Ontario Round Table on Environment and Economy.* At: <http://www.law.ntu.edu.tw/sustain/intro/ortee/20/21making.html>.

Rickson, R & Burch, D (1996) Contract farming in organizational agriculture: The effects upon farmers and the environment. In D Burch, R Rickson & G Lawrence (eds.) *Globalization and Agri-food Restructuring: Perspectives from the Australasia Region.* Avebury, London, pp. 173–202.

Ritchie, John (ed.) (1971) *The Evidence to the Bigge Reports.* Volume 2. Heinemann, Melbourne.

Robbins, J (1981) Community development in Nuriootpa: The search for a model. In M Bowman (ed.) *Beyond the City: Case Studies in Community Structure and Development.* Longman Cheshire, Melbourne, pp. 144–65.

Robertson, G (1997) Managing the environment for profit. In Australian Bureau of Agricultural and Resources Economics (ABARE) *Outlook 97, Volume 2 Agriculture, Proceedings of the National Agricultural and Resources Outlook Conference.* ABARE, Canberra, pp. 75–79.

Robinson, G (2002) Nature, society and sustainability. In I Bowler, C Bryant & C Cocklin (eds.) *The Sustainability of Rural Systems: Geographical Interpretations.* Kluwer Academic, Dordrecht, pp. 35–57.

Rogers, M (1997) Food markets – capturing emerging opportunities. Australian Bureau of Agricultural and Resources Economics (ABARE) *Outlook 97, Volume 2 Agriculture, Proceedings of the National Agricultural and Resources Outlook Conference.* Canberra, ABARE, pp. 27–37.

Rogers, M & Collins, Y (eds.) (2001) *The Future of Australia's Country Towns.* Centre for Sustainable Regional Communities, La Trobe University, Bendigo.

Rolley, F & Humphreys, J (1993) Rural welfare: The human face of Australia's countryside. In A Sorensen & R Epps (eds.) *Prospects and Policies for Rural Australia.* Longman Cheshire, Melbourne, pp. 241–57.

Rose, J (1966) Dissent from down under: Metropolitan primacy as the normal state. *Pacific Viewpoint* 7: 1–27.

Rose, N (1996) The death of the social? Re-figuring the territory of government. *Economy and Society* 25(3): 327–56.

Ross, R & Trachte, K (1990) *Global Capitalism: The New Leviathan.* SUNY Press, Albany.

Rowland, DT (1982) Urbanization and internal migration. In ESCAP *Population of Australia*. Volume 1, Country Monograph Series No. 9. United Nations, New York, pp. 71–100.

Royal Commission on Water Supply (1885) Irrigation in Western America, First Report of Royal Commission on Water Supply. *Victorian Parliamentary Papers 2*. Quoted La Nauze, JA (1965) *Alfred Deakin: A Biography*. Melbourne University Press, Melbourne.

Rural Industries Research and Development Corporation (RIRDC) (1994) *Asian Food: Getting a Bigger Bite*. Research Paper No 94/5. RIRDC, Canberra.

Ruspini, E (2000) Women and poverty: A new research methodology. In D Gordon & P Townsend (eds.) *Breadline Europe: The Measurement of Poverty*. Policy Press, Bristol, pp. 107–40.

Ruthven, P (2001) Where is the dairy industry heading? *Proceedings, Gippsland Dairy Conference*. Gippsland Dairy Industry Committee.

Salt, B (1992) *Population Movements in Non-metropolitan Australia*. AGPS, Canberra.

—— (1999) *Population Growth*. Tenth edition. KPMG, Melbourne.

—— (2001) *The Big Shift: Welcome to the Third Australian Culture*. Hardie Grant Books, South Yarra.

Sammon, M & Thomson, M (2003) *Private Investor Needs for Land Stewardship Investment*. Department of Sustainability and Environment, Melbourne.

Sant, ME & Simons, PL (1993) The conceptual basis of counterurbanisation: Critique and development. *Australian Geographical Studies* 31(2): 113–26.

Santamaria, BA (1943) Submission and Evidence to Rural Reconstruction Commission 18 October 1943. Qs 10825–58, CP 462/1, Bundle 4, National Archives of Australia.

Schirato, A & Danaher, G (2002) *Understanding Bourdieu*. Allen & Unwin, Sydney.

Schnaiberg, A (1980) *The Environment: From Surplus to Scarcity*. Oxford University Press, New York.

Scoones, I (1998) *Sustainable Rural Livelihoods: A Framework for Analysis*. Institute of Development Studies, Brighton.

Scott, K, Park, J & Cocklin, C (2000) From 'sustainable rural communities' to 'social sustainability': Giving voice to diversity in Mangakahia Valley, New Zealand. *Journal of Rural Studies* 16(4): 433–46.

Sefton, JA & Weale, MR (1996) The net national product and exhaustible resources: The effects of foreign trade. *Journal of Public Economics* 61: 21–47.

Select Committee on Agricultural Industry ... dealing with Health and Hygiene in the Country (1920) Fifth Interim Report of the Select Committee on Agricultural Industry ... dealing with Health and Hygiene in the Country. *New South Wales Parliamentary Papers* 1.

Select Committee on the Disposal of Lands in the Colonies (1836) Evidence of Thomas Malthus to the Select Committee on the Disposal of Lands in the Colonies. *British Parliamentary Papers*. Irish University Press Reprint Series, Colonies, General, 2.

Share, P, Campbell, H & Lawrence, G (1991) The vertical and horizontal restructuring of rural regions. In M Alston (ed.) *Family Farming: Australia and New Zealand*. Centre for Rural Social Research, Charles Sturt University, Wagga Wagga.

Sheehan, P & Tegart, G (eds.) (1998) *Working for the Future: Technology and Employment in the Global Knowledge Economy*. Victoria University Press for the Centre for Strategic Economic Studies, Melbourne.

Shortall, S (2002) Irish farm women. Paper presented to the Setting the Agenda for Rural Women Conference. Centre for Rural Social Research, Charles Sturt University, Wagga Wagga.

Shucksmith, M (2000) *Exclusive Countryside? Social Exclusion and Regeneration in Rural Britain*. Joseph Rowntree Foundation, York. At: <http://www.jrf.org.uk/bookshop/eBooks/1859351271.pdf>.

Smailes, P (1996) Entrenched farm indebtedness and the process of agrarian change: A case study and its implications. In D Burch, R Rickson & G Lawrence (eds.) *Globalization and Agri-food Restructuring: Perspectives from the Australasia Region*. Avebury, London, pp. 301–22.

—— (2002) From rural dilution to multifunctional countryside: Some pointers to the future from South Australia. *Australian Geographer* 33: 79–95.

Smailes, P, Argent, N & Griffin, T (2002a) Rural population density: Its impacts on social and demographic aspects of rural communities. *Journal of Rural Studies* 18(4): 385–404.

Smailes, P, Argent, N, Griffin, T & Mason, G (2002b) *Rural Community Social Area Identification*. National Centre for Social Applications of Geographical Information Systems, Adelaide.

Smailes, P & Hugo, G (2003) The Gilbert Valley, South Australia. In C Cocklin & M Alston (eds.) *Community Sustainability in Rural Australia: A Question of Capital?* Centre for Rural Social Research, Charles Sturt University, Wagga Wagga, pp. 65–106.

Smith, B (1998) Participation without power: Subterfuge or development? *Community Development Journal* 33(3): 197–204.

Smith, CL & Steel, BS (1995) Core-periphery relationships of resource-based communities. *Journal of the Community Development Society* 26(1): 52–70.

Sneddon, CS (2000) 'Sustainability' in ecological economics, ecology and livelihoods. *Progress in Human Geography* 24: 521–49.

Sobels, J, Curtis, A & Lockie, S (2001) The role of Landcare group networks in rural Australia: Exploring the contribution of social capital. *Journal of Rural Studies* 17(3), 265–76.

Social Exclusion Unit (SEU) (2001) *Preventing Social Exclusion*. At: <http://www.socialexclusionunit.gov.uk/publications/reports/html/pse/pse_html/index.htm>.

Society of St Vincent de Paul (1998) *A Country Crisis: A Report on Issues Confronting Rural Victoria*. Society of St Vincent de Paul, Melbourne.

Solow, R (1974) Intergenerational equity and exhaustible resources. *Review of Economic Studies Symposium*: 29–45.

—— (1986) On the intergenerational allocation of natural resources. *Scandinavian Journal of Economics* 88: 141–49.

—— (1993) An almost practical step towards sustainability. *Resources Policy* 19: 162–72.

Sorensen, A (1993) The future of the country town: Strategies for local economic development. In A Sorensen & R Epps (eds.) *Prospects and Policies for Rural Australia*. Longman Cheshire, Melbourne, pp. 201–40.

—— (2000) *Regional Development: Some Issues for Policy Makers*. Parliamentary Library Research Paper 26, Canberra: Parliament of Australia.

Sorensen, A & Epps, R (eds.) (1993) *Prospects and Policies for Rural Australia*. Longman Cheshire, Melbourne.

—— (1996) Community leadership and local development: Dimensions of leadership in four Central Queensland towns. *Journal of Rural Studies* 12(2): 113–25.

Spaling, H & Smit, B (1995) A conceptual model of cumulative environmental effects of agricultural land drainage. *Agriculture, Ecosystems and Environment* 53(2): 99–108.

Squires, V & Tow, P (eds.) (1991) *Dryland Farming: A Systems Approach.* Sydney University Press, Sydney.

Stacey, M (1969) The myth of community studies. *The British Journal of Sociology* 20(2): 134–47.

Staples, M & Millmas, A (eds.) (1998) *Studies in Australian Rural Economic Development.* Centre for Rural Social Research, Charles Sturt University, Wagga Wagga.

Stayner, R (1996) Policy issues in farm adjustment. In A Burdon (ed.) *Australian Rural Policy Papers,* 1990–1995. AGPS, Canberra.

—— (2003) Guyra, New South Wales. In C Cocklin & M Alston (eds.) Community *Sustainability in Rural Australia: A Question of Capital?* Centre for Rural Social Research, Charles Sturt University, Wagga Wagga, pp. 38–64.

Stayner, R & Reeve, I (1990) *Uncoupling: Relationships between Agriculture and the Local Economies in Rural Areas of New South Wales.* Rural Development Centre, University of New England, Armidale.

Statham, Pamela (1989) Patterns and perspectives. In Statham, Pamela (ed.) *The Origins of Australia's Capital Cities.* Oxford University Press, Melbourne, pp. 1–36.

Stevens, A, Bur, A & Young, L (2003) People, jobs, rights and power: The roles of participation in combating social exclusion in Europe. *Community Development Journal* 38(2): 84–95.

Stimson, R (2001) Dividing societies: The socio-political spatial implications of restructuring in Australia. *Australian Geographical Studies* 39: 198–216.

Sustainable Community Roundtable (1999) *Sustainable Community in South Puget Sound – 1999 Update.* Sustainable Community Roundtable, Olympia, WA.

Sustainable Land and Water Resources Management Committee Working Group on Dryland Salinity (2000) *Management of Dryland Salinity: Future Strategic Directions in the Context of Developing a National Policy for Natural Resource Management.* CSIRO Publishing, Melbourne.

Suttles, G (1972) *The Social Construction of Communities.* Chicago University Press, Chicago.

Swartz, D (1997) *Culture and Power: The Sociology of Pierre Bourdieu.* University of Chicago Press, Chicago and London.

Taylor, M (1991) Economic restructuring and regional change in Australia. *Australian Geographical Studies* 29(2): 255–67.

Taylor, TG (1951) *Australia: A Study of Warm Environments and their Effect on British Settlement.* Methuen, London.

Teather, E (1992) Remote rural women's ideologies, spaces and networks: Country Women's Association of New South Wales. *Australian and New Zealand Journal of Sociology* 28(3): 209–22.

Thomashow, M (1995) *Ecological Identity: Becoming a Reflective Environmentalist.* MIT Press, Massachusetts.

Tilly, C (1974) Do communities act? In MP Effratt (ed.) *The Community: Approaches and Applications.* The Free Press, New York, pp. 209–40.

Tonts, M (2000) The restructuring of Australia's rural communities. In W Pritchard & P McManus (eds.) *Land of Discontent: The Dynamics of Change in Rural and Regional Australia.* UNSW Press, Sydney, pp. 52–72.

Tonts, M & Black, A (2003) Narrogin, Western Australia. In C Cocklin & M Alston (eds.) *Community Sustainability in Rural Australia: A Question of Capital?* Centre for Rural Social Research, Charles Sturt University, Wagga Wagga, pp. 107–34.

Tonts, M & Jones, R (1997) From state paternalism to neoliberalism in Australian rural policy: Perspectives from the Western Australian wheatbelt. *Space and Polity* 1(2): 171–90.

Townsend, P (1996) *A Poor Future: Can We Counter Growing Poverty in Britain and Across the World?* Lemos & Crane in association with the Friendship Group, London.

Townsend, P & Gordon, D (2000) Introduction: The measurement of poverty in Europe. In D Gordon & P Townsend (eds.) *Breadline Europe: The Measurement of Poverty.* Policy Press, Bristol, pp. 1–24.

Troughton, M (1995) Presidential address: Rural Canada and Canadian rural geography – an appraisal. *The Canadian Geographer* 39(4): 290–305.

—— (2002) Enterprises and commodity chains. In I Bowler, C Bryant & C Cocklin (eds.) *The Sustainability of Rural Systems: Geographical Interpretations.* Kluwer Academic Publishers, Dordrecht, pp. 123–45.

United Nations Conference on Environment and Development (1992) *Agenda 21.* At: <http://www.un.org/esa/sustdev/documents/agenda21/english/agenda21toc.htm>.

United Nations Development Project (UNDP) (1997) Capacity Development. Technical Advisory Paper 2. Bureau for Policy Development, UNDP Management and Governance Division, New York.

Uphoff, N (2000) Understanding social capital: Learning from the analysis of experience of participation. In I Serageldin & P Dasgupta (eds.) *Social Capital: A Multifaceted Perspective.* World Bank, Washington, pp. 215–49.

Vance, J (1970) *The Merchant's World: The Geography of Wholesaling.* Prentice-Hall, Englewood Cliffs, NJ.

Vanclay, F & Lawrence, G (1995) *The Environmental Imperative: Ecosocial Concerns for Australian Agriculture.* Central Queensland University Press, Rockhampton.

Van der Perk, J & de Groot, R (2000) *Towards a Method to Estimate Critical Natural Capital.* Working Paper 5, Department of Environmental Sciences, Wageningen University.

Venning, J & Higgins, J (eds.) (2001) *Towards Sustainability: Emerging Systems for Informing Sustainable Development.* UNSW Press, Sydney.

Victorian Catchment Management Council (VCMC) & Department of Sustainability and Environment (DSE) (2003) *Ecosystem Services through Land Stewardship Practices: Issues and Options.* DSE, Melbourne.

Victorian Legislative Assembly Select Committee on the Drift of Population from the Country Districts to the City (1918) Report of Victorian Legislative Assembly Select Committee on the Drift of Population from the Country Districts to the City. *Victorian Parliamentary Papers* 1(D1).

Vidich, AJ & Bensman, J (1960) *Small Town in Mass Society.* Doubleday Anchor, New York.

Vinson, A (1999) *Unequal in Life: The Distribution of Social Disadvantage in Victoria and New South Wales.* Jesuit Social Services, Richmond, Victoria.

Wadham, S, Wilson, R & Wood, J (1964) *Land Utilization in Australia.* Melbourne University Press, Parkville.

Wadham, SM (1943a) *The Land and the Nation.* Stockland Press, Melbourne.

—— (1943b) *Reconstruction and the Primary Industries.* Melbourne University Press, Melbourne.

Wahlquist, A (1997) Series on rural decline, *The Land.* Australian Rural Press, Richmond.

Wakefield, Edward Gibbon (1929) *Letter from Sydney and Other Writings.* Everyman edition, JM Dent, London. (Originally published 1829.)

Walker, A (1997) Introduction: The strategy of inequality. In A Walker & C Walker (eds) *Britain Divided: The Growth of Social Exclusion in the 1980s and 1990s*. CPAG, London, pp. 1–16.

Walker, A & Walker, C (1997) Conclusion: Prioritise poverty now. In A Walker & C Walker (eds.) *Britain Divided: The Growth of Social Exclusion in the 1980s and 1990s*. CPAG, London, pp. 279–88.

Wall, M & Cocklin, C (1996) Attitudes towards forestry in the East Coast region. *New Zealand Forestry* 41: 21–27.

Warburton, D. (1998) A passionate dialogue: Community and sustainable development. In D Warburton (ed.) *Community and Sustainable Development: Participation in the Future*. Earthscan, London, pp. 1–39.

Ward, G (1996) Population growth in south-east Queensland. In P Newton & M Bell (eds.) *Population Shift: Mobility and Change in Australia*. AGPS, Canberra, pp. 165–181.

Weekly Times (1918). 15 June.

Wentworth Group of Concerned Scientists (2003) *A New Model for Landscape Conservation in New South Wales*. Report submitted to Premier Carr. At: <http://www.wwf.org.au/News_and_information/Publications/PDF/Report/new_model_report_to_carr.pdf>

Western Australia Department of Commerce and Trade (1999) *Capacity Building in Regional Western Australia: A Regional Development Policy for Western Australia*. Technical Paper. Western Australia Department of Commerce and Trade, Perth.

Westgarth, William (1854) *Victoria and the Australian Goldmines*. London. Quoted in A Dingle, *Settling*. Fairfax, Syme & Weldon, McMahon's Point.

Whatmore, S (1991) *Farming Women: Gender, Work and Family Enterprise*. Macmillan, London.

Whatmore, S, Marsden, T & Lowe, P (1994) Introduction: Feminist perspectives in rural studies. In S Whatmore, T Marsden & P Lowe (eds.) *Gender and Rurality*. David Fulton Publishers, London, pp. 1–10.

Whittaker, A & Banwell, C (2002) Positioning policy: The epistemology of social capital and its application in applied rural research in Australia. *Human Organization* 61(3): 252–61.

Whittenbury, K (2003) *Exploring the Taken-for-granted: Cultural and Structural Influences on Men's and Women's Perceptions of Family Life in a Rural Community*. Unpublished PhD thesis, Charles Sturt University, Wagga Wagga.

Wild, RA (1978) *Bradstow: A Study of Status, Class and Power in a Small Australian Town*. Revised edition. Angus & Robertson, Sydney.

—— (1983) *Heathcote*. George Allen & Unwin, Sydney.

Wilkinson, J, Gray, I & Alston, M (2003) Tumbarumba, New South Wales. In C Cocklin & M Alston (eds.) *Community Sustainability in Rural Australia: A Question of Capital?* Centre for Rural Social Research, Charles Sturt University, Wagga Wagga, pp. 135–69.

Williams, R (2003) Recognising and valuing women in developing policy for community capacity building: A focus on the interplay between leadership and social capital. In R O'Hagan, M Alston & S Spriggs (eds.) *Setting the Agenda for Rural Women: Research Directions*. Conference Proceedings and Recommendations. Centre for Rural Social Research, Charles Sturt University, Wagga Wagga, pp. 161–76.

Williams, Raymond (1975) *The Country and the City*. Paladin, St Albans UK. (First published 1973.)

Wilson, G (2001) From productivism to post-productivism ... and back again?

Exploring the (un)changed natural and mental landscapes of European agriculture. *Transactions of the Institute of British Geographers* 26(1): 77–102.

Wiseman, J (1998) *Global Nation? Australia and the Politics of Globalisation*. Cambridge University Press, Cambridge.

Withagen, C (1996) Sustainability and investment rules. *Economics Letters* 53(1): 1–6.

Wolf, S & Wood, S (1997) Precision farming: Environmental legitimation, commodification of information, and industrial coordination. *Rural Sociology* 62: 180–206.

Woolcock, M & Narayan, D (2000) Social capital: Implications for development theory, research, and policy. *World Bank Research Observer* 15(1): 225–49.

World Commission on Environment and Development (1987) *Our Common Future*. Oxford University Press, Oxford.

World Conservation Union (IUCN), United Nations Environment Programme (UNEP) & World Wide Fund For Nature (WWF) (1991) *Caring for the Earth: A Strategy for Sustainable Living*. IUCN, UNEP, WWF, Gland, Switzerland.

Worthington, AC & Dollery, BE (2000) Can Australian local government play a meaningful role in the development of social capital in disadvantaged rural communities? *Australian Journal of Social Issues* 35(4): 349–61.

Wrigley, HN (1988) *The Narraburra Cup*. Soil Conservation Service of New South Wales/Temora Shire Council, Temora.

Wyn, J, Stokes, H & Stafford, J (1998). *Young People Living in Rural Australia in the 1990s*. Youth Research Centre, Faculty of Education, University of Melbourne, Parkville, Victoria.

Wynen, E (2002) Multifunctionality and Agriculture – Why the fuss? *Current Issues Brief* No. 13, Parliamentary Library Information and Research Services, Canberra.

Yenken, D & Porter, L (2001) *A Just and Sustainable Australia*. Australian Council of Social Service, on behalf of The Australian Collaboration, Redfern and Melbourne.

Yencken, D & Wilkinson, D (2000) *Resetting the Compass: Australia's Journey Towards Sustainability*. CSIRO Publishing, Melbourne.

Young, M (1997) Mining or minding: Opportunities for Australia to improve conservation of remnant vegetation and to alleviate land degradation. In Environment Australia, *Proceedings, Environmental Economics Round Table*. Environment Australia, Canberra.

Young, M & McColl, J (2003) Robust separation: A search for a generic framework to simplify registration and trading of interests in natural resources. In Land and Water Australia, *Property: Rights and Responsibilities – Current Australian Thinking*. Land and Water Australia, Canberra, pp. 55–70.

Young, M, Shi, T & Crosthwaite, J (2003) *Duty of Care: An Instrument for Increasing the Effectiveness of Catchment Management*. Department of Sustainability and Environment, Melbourne.

Zablocki, BD (2000) What can the study of communities tell us about communities? In EW Lehman (ed.) *Autonomy and Order: A Communitarian Anthology*. Rowman and Littlefield, Lanham, pp. 71–88.

INDEX